Particle Technology Series

Volume 26

Series editor

José Manuel Valverde Millán, University of Sevilla, Spain

Many materials exist in the form of a disperse system, for example powders, pastes, slurries, emulsions and aerosols, with size ranging from granular all the way down to the nanoscale. The study of such systems necessarily underlies many technologies/products and it can be regarded as a separate subject concerned with the manufacture, characterization and manipulation of such systems. The series does not aspire to define and confine the subject without duplication, but rather to provide a good home for any book which has a contribution to make to the record of both the theory and applications of the subject. We hope that engineers and scientists who concern themselves with disperse systems will use these books and that those who become expert will contribute further to the series.

The Springer Particle Technology Series is a continuation of the Kluwer Particle Technology Series, and the successor to the Chapman & Hall Powder Technology Series.

More information about this series at http://www.springer.com/series/6433

John G. Yates · Paola Lettieri

Fluidized-Bed Reactors: Processes and Operating Conditions

Springer

John G. Yates
Department of Chemical Engineering
University College London
London
UK

Paola Lettieri
Department of Chemical Engineering
University College London
London
UK

ISSN 1567-827X
Particle Technology Series
ISBN 978-3-319-81919-8 ISBN 978-3-319-39593-7 (eBook)
DOI 10.1007/978-3-319-39593-7

Printed on acid-free paper

This Springer imprint is published by Springer Nature
The registered company is Springer International Publishing AG Switzerland

This book is dedicated to the next generation; Alice, Robin, Polly, Georgia, Phoebe and Jasper

Foreword

Fluidized beds are ubiquitous in a variety of chemical and physical processing applications as well as in natural phenomena. Despite many decades of intense research efforts on fundamentals and applications, fluidized beds still disclose unrevealed features and pose challenging issues to the researcher, the process engineer, the natural scientist.

Fluidized-Bed Reactors: Processes and Operating Conditions offers the reader an up-to-date survey of successful applications of the fluidized bed technology to chemical and process industry. The book is unique in that it provides a historical perspective on the development of fluidized beds industrial applications, addressing both the success stories and the operational problems encountered in developing fluidized bed processes. This "evolutionary" approach is helpful in developing a rationale for the use of fluidized beds and in discussing the proper selection of process conditions for successful design and scale-up and for trouble-free operation of fluidized bed process units.

Of particular interest is the consideration of the influence of temperature and pressure on fluidization in many industrial processes operated at far-from-ambient conditions. Moreover the importance of the often-disregarded role of interparticle forces is scrutinized and surveyed. This analysis benefits from the well-recognized competence of the authors on fluidization under "extreme" process conditions and on fluidization of fine particles.

The book is an excellent and instructive reading for undergraduate and graduate students in chemical engineering and industrial chemistry and provides useful information to professional process engineers and chemists.

Piero Salatino

Preface

Fluidization, a technique in which an assembly of solid particles is held in suspension by an upward-flowing fluid, has been practised in the process industries for over 80 years. An early application was the Winkler process for the gasification of coal closely followed by the fluidized catalytic cracking process for the production of gasoline. Since the middle decades of the twentieth century the technique has spread widely from fine chemicals to the heavy industries such as uranium processing and sulfide-ore roasting so that today it represents along with distillation, crystallization, filtration, etc. one of the major unit operations in the chemical engineer's toolbox. Running in parallel with these industrial developments has been a truly enormous research and development activity in both industry and academia aimed at achieving a better understanding of the hydrodynamics of fluidized beds and at how this may be applied to the improvement of existing processes and the development of new ones.

The focus in the present volume is on the design and operation of the centrepiece of a process, the reactor itself, emphasizing the reasons for choosing fluidized beds rather than alternatives such as fixed beds in the context of a particular process. The book begins with a brief history of fluidization covering decade by decade the period from the 1940s to the first decade of the twenty-first century describing the processes introduced and highlighting their successes and failures. Basic research into the subject is outlined and attention drawn to the theoretical and experimental advancements achieved during this period. The second chapter considers industrial processes based on heterogeneous catalysis such as olefin polymerization and catalytic cracking while the succeeding chapter looks at the non-catalytic routes to materials such as ultra-pure silicon and titanium dioxide; fluidized-bed combustion and gasification of conventional fuels are also considered as well as the relatively new technique of chemical looping combustion. Chapter 4 is concerned with conversion processes for biomass and waste materials while Chap. 5 looks at the effects of temperature, pressure and particle-size distribution on design and operating conditions. The book concludes with a chapter on the scaling and scale-up of reactors.

The treatment throughout is at the postgraduate/postexperience level but practising engineers and scientists concerned with gas–solid systems will, it is hoped, find much of interest.

London, UK
2016

John G. Yates
Paola Lettieri

Acknowledgements

The authors wish to acknowledge those who have been so helpful in preparing the material for this book. We are very grateful to the Fluidization Group in Chemical Engineering at University College London, in particular to Dr. Massimiliano Materazzi for his contribution in writing Chap. 4, Mr. Domenico Macri' for his assistance with sorting out the copyright for all the figures included in the book and to Miss Carla Tagliaferri for assisting in the literature search. We are also grateful to Prof. Gregory Patience of the Ecole Polytechnique Montreal for helpful correspondence relating to the DuPont CFB process for maleic anhydride.

We have dedicated Chap. 5 to our dear friend and colleague, the late Dr. David Newton, for his contribution to the research described in the chapter. David was an expert and a practitioner in the field of fluidization. He had joined John Yates' research group at UCL in 1981 to work on the effect of fine particles on the catalytic dehydrogenation of butene to butadiene, a model reaction used to probe the basic fluid dynamics of these reactors. David went from UCL directly to the BP Research Centre at Sunbury-on-Thames, UK where he continued his work in fluidization and became Head of the Fluidization Group at BP Chemicals. Paola joined David's Group at BP in 1995 and together with John continued to work with David on the many challenges described in Chap. 5 until December 2009, when David sadly passed away after being diagnosed with asbestosis.

Finally a word of sincere thanks to the staff at Springer, particularly Mieke van der Fluit, for their constant encouragement and guidance.

Contents

Notations

A Bed area (m^2)
Ar Archimedes number, $\frac{d_p^3\rho_g(\rho_p-\rho_g)g}{\mu^2}$ (–)
C_A Concentration of reactant A (mol/m^3)
d_p Particle diameter (m)
ds Surface diameter (m)
dv Volume diameter (m)
D Bed diameter (m)
f_b Bubble volume fraction (–)
Fa Attraction force between particles (kg m/s^2)
Fr Froude number (–)
F_{45} Size fraction <45μm (–)
g Acceleration due to gravity (m/s^2)
Ga Galileo number (–)
G_s Solids mass flux (kg/m^2s)
h Heat-transfer coefficient (W/m^2K)
k Reaction rate coefficient (s^{-1})
K Interphase mass-transfer coefficient (s^{-1})
L Bed height (m)
n Richardson–Zaki index (–)
p Fluid pressure (N/m^2)
q Heat-transfer rate (W)
Q Volumetric gas flowrate (m^3/s)
r Reaction rate (mol/s)
R Particle radius (m)
Re Reynolds number, $\frac{ud_p\rho_g}{\mu}$ (–)
T Temperature (K)
u Fluid velocity (m/s)
u_D Dynamic wave velocity (m/s)
u_K Kinematic wave velocity (m/s)

u_t Particle terminal-fall velocity (m/s)
U Superficial fluid velocity (m/s)
W Dimensionless rupture energy (–)
ε Void fraction (–)
ρ Density (kg/m^3)
μ Dynamic viscosity (Ns/m^2)
v Kinematic viscosity (m^2/s)
τ Gas residence time (s)
ϕ Particle sphericity (–)

Subscripts

bc Bubble to cloud
ce Cloud to emulsion
f Fluid
g Gas
m_b Minimum bubbling
m_f Minimum fluidization
p Particle
s Solid

Chapter 1
Introduction

Abstract This introductory chapter begins with a description of the various flow regimes observed in beds of solid particles fluidized with gases or liquids. Distinctions are drawn between the gas-fluidized behaviour of fine, low density particles, Group A in the Geldart classification, larger, more dense materials, Group B, and those more difficult to fluidize at all, Groups C and D. It is shown how gas-fluidized beds behave as the velocity of the gas flowing through them is increased the transitions from bubbling to slugging to turbulent to transport behaviour being described and illustrated by means of schematic diagrams. There then follows a discussion of the historical development of fluidization from its first major industrial application in the 1940s through subsequent decades up to the first decade of the new millennium. Theoretical advances are described from the early "two-phase" theory of Toomey and Johnstone through Davidson's analysis of the flow of gas through bubbling beds to the more recent "particle-bed" model. Industrial applications of gas-fluidized-bed reactors are discussed with reference to fluidized catalytic cracking, naphthalene oxidation, coal to gasoline via the Synthol process, propylene ammoxidation to acrylonitrile, ethylene polymerisation and butane oxidation to maleic anhydride. The wide range of experimental techniques used in basic fluidization research is described as well as the recent application of computational fluid dynamics to gas-solid fluidized systems.

1.1 Regimes of Fluidization

Two-phase systems: liquid-solid; gas-solid

An assembly of solid particles becomes fluidized when the combined drag and body forces exerted on the particles by an upward-flowing fluid exceed the gravitational force holding the assembly together. At this point the particles are suspended in the fluid and are free to move. The fluid velocity at which this occurs is termed the "minimum fluidization velocity" U_{mf} and its value is a function of the size, shape and density of the particles and the density and viscosity of the fluid. When the velocity

J.G. Yates and P. Lettieri, *Fluidized-Bed Reactors: Processes and Operating Conditions*, Particle Technology Series 26,
DOI 10.1007/978-3-319-39593-7_1

of the fluid is increased above U_{mf} the behaviour of the assembly, known as a "fluidized bed", depends on the nature of both fluid and particle. If the fluid is a liquid the bed of particles expands more or less uniformly the particles moving about freely in the fluid flow field and stabilising at a void fraction, ε, which is a function of the fluid velocity, U, the terminal fall velocity of the particles in the fluid, U_t, and an exponent, n, which is a function of the fluid-particle system in question; n correlates with the terminal Reynolds number, Re_t, but is constant in the viscous and inertial flow regimes at 4.8 and 2.4 respectively. Thus:

$$U/U_t = \varepsilon^n \tag{1.1}$$

Equation (1.1) is known as the Richardson-Zaki equation and applies to a wide range of particle-liquid combinations (see below).

If the fluidizing fluid is a gas the behaviour of the bed as U exceeds U_{mf} depends on the nature of the particles. Geldart (1973) classified fluidizable particles into four groups, A, B, C and D as shown in Fig. 1.1.

As U_{mf} is exceeded particles in Group A expand uniformly in the same way as a liquid-fluidized bed up to a gas velocity at which the bed collapses and voids called "bubbles" begin to form; these rise through the bed at a velocity proportional to their size and burst at the bed surface. The gas velocity at which this occurs is termed the "minimum bubbling velocity, U_{mb}". A typical example of a Group A powder is cracking catalyst used in the FCC process to be described later. Group B particles such as coarse sand begin to form bubbles immediately U exceeds U_{mf} a type of behaviour referred to as "aggregative fluidization". The particles in Group C, typically flour and cement, tend to be cohesive and are difficult to fluidize at all whereas the relatively large and dense materials in Group D such as lead shot are prone to spouting rather than fluidizing. The great majority of fluidized-bed reactors use powders in Groups A and B.

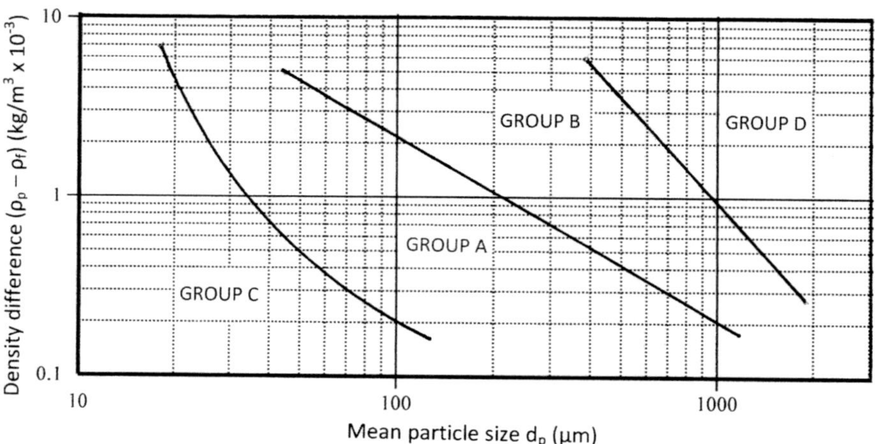

Fig. 1.1 The Geldart classification of powders

Once the bubbling regime has been established in beds of A and B materials any increase in volumetric gas flow rate causes the bubbles to coalesce, grow in size and increase in velocity as they pass to the bed surface. With increasing superficial gas velocity in beds of material contained in a vessel with a high aspect ratio (height to diameter ratio) a point is reached at which the bubble diameter becomes equal to that of the container and the bed is then said to be "slugging" and is characterised by the periodic passing of large bubbles or slugs and large fluctuations in bed pressure drop corresponding to the slug frequency.

In bubbling beds of wide diameter particles are ejected into the freeboard space above the bed surface by the bursting bubbles the amount of material entrained increasing as the gas velocity is increased. At sufficiently high gas velocities bubble flow breaks down, the bed becomes highly turbulent and entrained particles must be separated from the gas flow via a cyclone separator and returned to the bed. The transition from bubbling to turbulent has been characterised by a critical gas velocity, $U_{c,}$ at which the standard deviation of pressure fluctuations measured at the base of the bed reaches a maximum value and axial and radial solids density profiles are more or less uniform over time. Operating gas velocities in the turbulent regime are in the range 0.4–1.2 ms^{-1}. A further transition from turbulent to so-called "fast" fluidization occurs at gas velocities in the range 1.2–15 ms^{-1}. At these velocities considerable entrainment of particles occurs and the bed develops a lean-core-annulus structure with a parabolic radial solids density profile. Solids are observed to move downwards at the walls of the containing vessel in a refluxing type motion. To maintain steady-state operation solids circulation via an external standpipe is necessary and as a result the beds are often turbulent in the base region and fast at the top with a pronounced axial solids density profile. These systems are normally referred to as circulating fluidized beds or CFB's.

A schematic diagram of the various regimes of fluidization is shown in Fig. 1.2 and a detailed discussion of much of the work carried out to distinguish between them is given by Grace and Bi (1997).

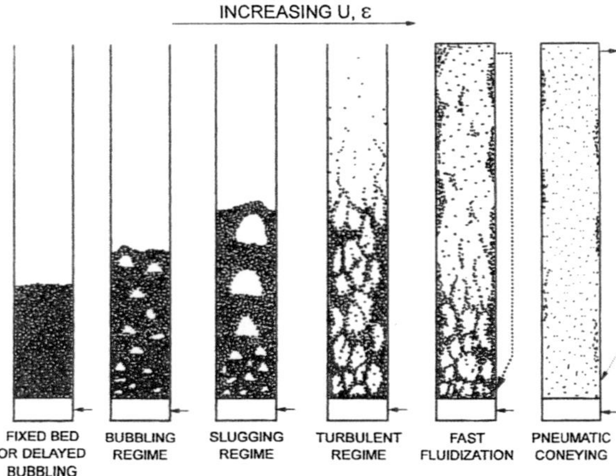

Fig. 1.2 Flow patterns in gas-solid fluidized beds (Grace and Bi 1997)

Three-phase systems: gas-liquid-solid

In these systems solid particles in a column are held in suspension, i.e. fluidized, by an up-flowing liquid, a gas being introduced separately at the base of the column. Thus liquid and gas flow co-current upwards the liquid being the continuous phase and the gas the discontinuous, bubbling phase. Several different modes of operation are possible: the solids may be retained in the column as in the case of a conventional liquid-solid fluidized bed or they may be transported out at the top and reintroduced at the base as in the case of a fast gas-solid system. The hydrodynamics of three-phase fluidized beds are complex but have been reviewed comprehensively by Fan (1989, 2003). Three-phase systems have been applied in a number of important industrial processes such as the catalytic hydrogenation of heavy or residual oils, the Fischer-Tropsch process, fermentation and aerobic biological waste water treatment.

1.2 A Brief History of Fluidization

The following is a brief outline of the major highlights in the story of fluidization tracing the development of the subject from its emergence in the early 1940s to the present day and dividing the period up into decades. Like centuries decades are arbitrary divisions of time but they represent a convenient way of charting industrial and academic developments in this area. Furthermore there is no doubt that the period 1940–1950 was decisive in the story of fluidization.

1.2.1 The 1940s

1.2.1.1 Fluidized Catalytic Cracking

On 25 May 1942 the world's first commercial fluidized catalytic cracking unit began operation at the Baton Rouge refinery of the Standard Oil Company of New Jersey (Jahnig et al. 1980). There had been two industrial uses of fluidization prior to this but neither had been very successful and they were applied on a very limited scale. The fluidized catalytic cracking, or FCC, process was the first major application of the technique and its commercialisation was the result of an enormous research and development effort by oil companies, plant contractors and academics. The subsequent spread of the process to virtually every major oil refinery in the world is a measure of the success of that effort.

The driving force that led to the development of the process was the need for a gasoline of higher quality than had been available hitherto for the rapidly improving internal combustion engines that were coming onto the market in the 1920s and

30s. Gasoline had been produced since the earliest days of oil refining first as a straight-run distillate fraction and later by thermal cracking of heavier fractions such as gas oil. The quality of gasoline produced by thermal cracking was sufficiently superior to the straight-run variety as to make the process a viable proposition. The realisation of the link between the structure of the gasoline molecules and their performance in engines only came about in the 1920s and this led to the introduction of the "octane number" scale on which the linear n-heptane was given an octane number of zero and the branched iso-octane a number of 100. The performance of a gasoline was then compared with blends of these two materials in a standard test engine to determine the octane number of the fuel. The product from thermal cracking was more highly branched than the straight-run material and this accounted for its superior performance. Now it might be thought that a suitable catalyst could have been found that would increase the extent of branching even further. Catalytic processes had for long been used in the chemical industry for just such purposes but oil companies at that time were reluctant to become involved with catalytic chemistry. They preferred to keep to physical operations such as distillation or simple chemistry such as thermal decomposition with which they were familiar. The catalytic effect of aluminium trichloride on the cracking of oil fractions had in fact been observed in 1915 but it wasn't until the pioneering work of the French mechanical engineer Eugene Houdry in the mid-1930s that a catalytic cracking process became a reality. He discovered through trial and error experiments that a naturally occurring clay mineral that had been activated by treatment with acids was capable of carrying out the cracking reaction and of producing a highly branched material with a superior performance to the thermally cracked product.

Houdry discovered that in addition to producing gasoline his catalyst also produced a form of carbon called "coke" which was deposited on the catalyst surface and gradually reduced its activity to zero. He found however that the coke could be burned off by blowing air over the surface at about 500 °C. On the plant this was achieved by having three packed-bed reactors on stream simultaneously, one cracking, one regenerating and one being purged ready for the next stage. The whole cycle took 24 min and its success was strongly dependent on the use of high-temperature automatic valves and sophisticated control algorithms both of which were revolutionary at the time.

The evident success of the Houdry process caused the world's biggest oil company, Standard Oil New Jersey (now ExxonMobil), to become more interested in catalytic cracking than they had been hitherto so in early 1938 with a European war on the horizon and an anticipated need for large quantities of aviation gasoline the company took the decision to develop a catalytic cracking process that would be independent of Houdry's patents. A consortium, Catalytic Research Associates, was set up consisting of eight companies: Jersey Standard, Kellogg, Indiana Standard, Anglo Iranian (which later became BP), Universal Oil Products, Texaco, Shell and the German chemical giant I G Farben (which dropped out in 1940). As one observer has commented "This was a formidable grouping and, with 400 workers at Jersey and 600 in the other companies, represented probably the largest single concentration of scientific manpower in the world. It was also probably the greatest scientific effort directed at a single project and would be surpassed only by the development of the atomic bomb" (Enos 1962). The result of this effort was the FCC process. Early on in the development it had been decided to use a finely-powdered catalyst rather than the pelleted material used by Houdry. One advantage of this was that the catalyst could be moved continuously suspended in the oil vapour between the reactor and the regenerator thereby avoiding the need for intermittent operation. Experiments were carried out with so-called "snake" reactors in which oil vapour or air conveyed the catalyst through horizontal runs of pipe but the reactors were plagued with problems of plugging and high pressure drops. Warren K Lewis and Edward T Gilliland, academic chemical engineers based at MIT, were consultants to Standard Oil and came to the conclusion that overall vertical flow would be less susceptible to plugging than the snake reactors and would in addition give some contacting advantage due to the relative slip between the solid particles and the oil vapour. Following cold-flow experiments carried out at MIT a large-scale pilot plant was built on the lines suggested by Lewis and Gilliland and the operating feasibility of the process was demonstrated. One enormous advantage of this mode of operation is that not only are the catalyst particles regenerated and their activity restored but they are also heated by the exothermic combustion reaction in the regenerator and this heat is carried by the circulating particles into the reactor where it provides the thermal driving force for the endothermic cracking reaction. Further aspects of catalytic cracking will be discussed in Chap. 2.

1.2.1.2 Other Processes

Naphthalene oxidation

Another industrial process using fluidized-bed reactors and developed in this decade was that for the production of phthalic anhydride by the air oxidation of coal tar-derived naphthalene.

Phthalic anhydride is widely used by the chemical industry in the manufacture of dyestuffs, alkyd resins and plasticisers. Early processes for its production were based on the use of fixed-bed reactors packed with vanadium pentoxide catalyst but problems associated with their operation led to the introduction of fluidized bed-based systems first by the Sherwin Williams company in the USA. In a typical system the bed of catalyst was fluidized with air and liquid naphthalene was pumped in from storage through spray nozzles just above the air distributor. The reactor temperature, 340–380 °C, was controlled by using the exothermic heat of reaction to raise high-pressure steam through tubular heat exchangers immersed in the bed. With the decline of the coal industry in western countries from the 1960s onward naphthalene became progressively scarce and the alternative feedstock, o-xylene, was found to be unsuitable for use in fluidized-bed reactors. Furthermore improvements in the design of fixed-bed reactors using o-xylene feedstock led to the phasing out of naphthalene-based fluidized-bed systems the last such plant in the UK being shut down in 1972.

The Hydrocol process
In the mid-1940s attempts were made in the USA to use fluidization technology for the production of gasoline from synthesis gas, a mixture of carbon monoxide and hydrogen produced via Fischer-Tropsch chemistry. The result was the Hydrocol process operated for some years at a plant located at Brownsville, Texas. The reactor was scaled up from a 200 mm diameter pilot reactor to a full-scale 5 m diameter bubbling-bed unit that never performed to design capacity and was finally shut down in 1957 (Geldart 1967).

Sulfide Roasting
The fluidized-bed roasting of sulfide ores such as pyrite, FeS_2, to produce sulfur dioxide for sulfuric acid manufacture was developed in the 1950s by Dorr-Oliver in the USA, Badische Aniline (BASF) in Germany and Sumitomo in Japan. The exothermic oxidation reaction of the sulfide is carried out by fluidizing ore particles under 10 mm in size with air in large-diameter beds at temperatures in the range 650–1100 °C. Start-up of the units is normally carried out by gas firing and temperature control during operation is either by top spraying with water or heat exchange via immersed cooling coils. Large numbers of these systems have operated throughout the world and many of their reactor units are of considerable size the largest being said to be that built by Lurgi in Tasmania with a bed diameter of 12.5 m and a height of over 16 m (Kunii and Levenspiel 1991).

1.2.2 The 1950s

This period saw the start of much basic theoretical and experimental research into fluidization.

Experimental techniques
New experimental techniques for probing the hydrodynamics of fluidized beds were reported such as X-ray attenuation by Grohse (1955), optical sensors by Yasui and Johanson (1958) and capacitance probes by Morse and Ballou (1951).

Grose used X-rays to determine the instantaneous and time-averaged bulk density of a finely-divided powder fluidized at a range of gas velocities with different designs of gas distributor. The work demonstrated the two-phase nature of fluidized beds but gave little in the way of detailed structure. Yasui and Johanson used a small tungsten filament as light source coupled to a small mirrored glass prism cemented to one end of a 4 mm diameter quartz tube wrapped with aluminium foil. Two such probes separated by a short variable distance were positioned one above the other and the assembly immersed in a fluidized bed. When a bubble filled the space between the lamp and the prism light was transmitted to the prism and reflected out through the quartz tube into a vacuum phototube. Here the light pulse was converted into an electrical signal which was amplified and recorded on the moving chart of an oscillograph. The rise velocity of bubbles was estimated from the time lag between the signals from the two probes and their known distance apart. The results clearly showed that bubbles increase in size and velocity as they rise in the bed; increases in the particle size of the bed material and of the fluidizing gas velocity were also found to increase bubble rise velocity.

The solid particles used in fluidized beds are typically electrical insulators and the capacitance measured by an immersed two-plate probe will be a function of the concentration of solids between the plates. The principle was used Morse and Ballou to explore the two-phase nature of a number of fluidized-bed systems and although their results were largely qualitative their pioneering technique has been developed and improved by a number of groups in subsequent years [see for example Werther and Molerus (1973)].

Early studies of fluid-bed combustion were reported in this period by Yagi and Kunii (1955) and of fluid-bed heat transfer by Mickley and Fairbanks (1955).

A new theory

An important theoretical rationalisation was the "two-phase" theory proposed by Toomey and Johnstone (1952) which posited that all the gas in excess of that required to just fluidize the bed passed through in the form of bubbles. Thus if Q_0 is the total volumetric gas flow rate into the bed, Q_{mf} the minimum fluidization flow rate and Q_b the bubble flow rate:

$$Q_0 = Q_{mf} + Q_b \tag{1.2}$$

Dividing through by the bed area, A, gives the bubble flow in terms of the superficial velocities U and U_{mf}:

$$Q_b/A = U - U_{mf} \tag{1.3}$$

Subsequent experimental work however showed that in many systems the theory overestimates the visible bubble flow and a greater proportion of gas flows interstitially i.e. through the dense or emulsion phase (Grace and Clift 1974). It was suggested therefore that Eq. 1.3 be recast in the following form:

$$Q_b/A = U - U_{mf}(1 + nf_b) \tag{1.4}$$

where n is a positive number and f_b is the volume fraction of the bed occupied by bubbles. In the ideal two-phase theory of course $n = 0$ and the larger the value of n the larger is the emulsion-phase flow. Grace and Clift (1974) showed that measured deviations from the ideal two-phase theory vary over a wide range and that the value of n not only differs from one system to another but that within a given system it can vary with bed height and gas velocity.

The 1950s saw the beginning of attempts to model theoretically the performance of fluidized-bed reactors. Most of these early models were based on the two-phase theory as set out above with various assumptions being made regarding the form of reactant gas flow through the bed. Some assumed the gas to be either in plug flow or to be completely mixed (Shen and Johnstone 1955; Mathis and Watson 1956;

Lewis et al. 1959) while others assumed a form of dispersed flow (May 1959). Most included a term allowing for gas transfer between the bubble and emulsion phases. When tested against experimental observations however none of the models was able to account satisfactorily for the observed chemical conversions but these failures prompted the development of further and more sophisticated models in the succeeding decades. Bubbling-bed models have been reviewed by Yates (1983) and more recently by Ho (2003).

Liquid fluidization

As mentioned above beds of particles fluidized by liquids do not form bubbles (although there are exceptions in the case of very dense particles and light liquids) but expand in a regular manner with the bed voidage, ε, increasing as the liquid velocity increases. Richardson and Zaki (1954) carried out an extensive experimental investigation of liquid-solid systems. Most of the work was concerned with sedimentation but a number of experiments were carried out with a range of solids of different size and density fluidized with water. They plotted superficial liquid velocity, U, against bed voidage, ε, and found the linear relationship:

$$logU - nlog\varepsilon + logU_i \tag{1.5}$$

where n is the slope of the line and $logU_i$ is the intercept on the $logU$ axis corresponding to ε = 1. From Eq. (1.5) it follows that:

$$U/U_i = \varepsilon^n \tag{1.6}$$

where $U_i = U_t$ for sedimentation and:

$$logU_i = logU_{\mathrm{t}} - d/D \tag{1.7}$$

for fluidization where d/D is the particle-to-bed diameter ratio. For beds with low d/D values however U_i may be approximated to U_t leading to the Richardson-Zaki equation (1.1).

The Synthol process

Following some years of development work the Sasol company started up their Synthol process at the Sasolburg plant in South Africa in 1955. The technology was based on a combination of the Lurgi process which produces synthesis gas by the reaction of steam with coke and the Fischer-Tropsch process which converts the synthesis gas to a mixture of light oils and gasoline. Its success over the aforementioned Hydrocol process was due to the use of a circulating fluid-bed reactor rather than the dense-phase bubbling-bed system used at Brownsville. In the dilute riser section the reactant gases, hydrogen and carbon monoxide, carry the suspended powdered iron catalyst upwards at 3–12 m/s at temperatures of around 350 °C the product gases and catalyst then being separated in cyclones and the solids returned to the riser via a vertical standpipe. This oil-from-coal process has operated successfully for many years and has undergone many modifications and developments. It was of great benefit to the South African economy in the years when economic sanctions prevented the country from importing oil products from abroad.

1.2.3 The 1960s

The outstanding theoretical advance in this period was Davidson's analysis of bubble flow in gas-fluidized beds (Davidson 1961) soon verified experimentally by the group at the Atomic Energy Research Establishment, Harwell UK (Rowe 1964). The Davidson model was based on potential flow theory and made the following assumptions:

- bubble shape is circular in two-dimensional beds and spherical in three-dimensional beds
- the bed material behaves like an incompressible fluid of zero viscosity
- the fluidizing gas flows like an incompressible viscous fluid that obeys Darcy's law.

The theory allows gas streamlines around rising bubbles to be calculated (Fig. 1.3).

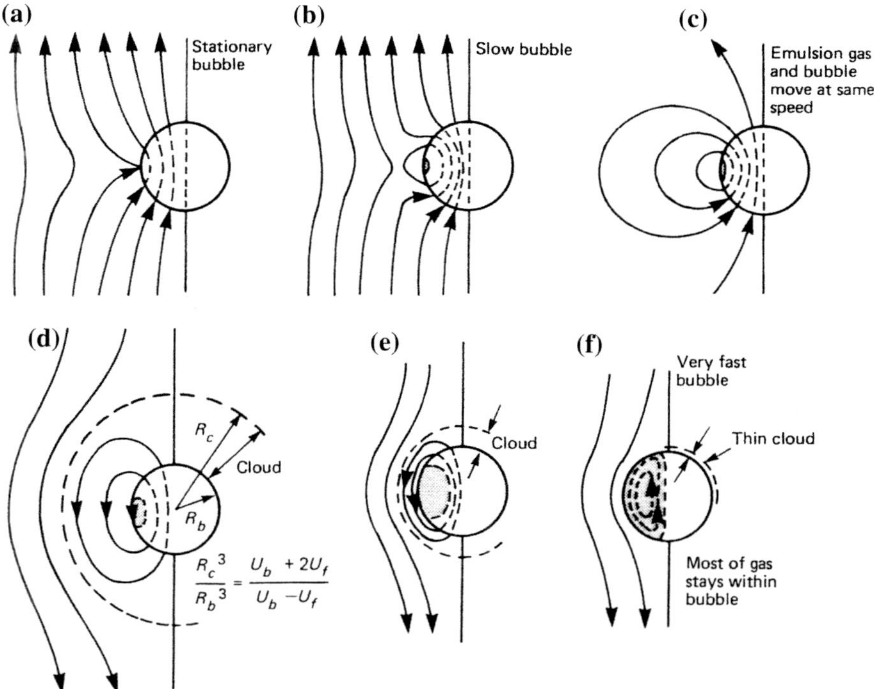

Fig. 1.3 Streamlines of gas near a single rising bubble from the Davidson model (Kunii and Levenspiel 1991)

The form of the streamlines is critically dependent on the ratio of bubble velocity, U_b, to interstitial gas velocity, U_f. When $U_b < U_f$, as commonly occurs in beds of large particles, (Fig. 1.3b), gas flows in at the base of the bubble and out at the top although some circulates in an annular ring between emulsion and bubble. In beds of small particles (Group A) bubbles normally rise faster than interstitial gas and for this case Davidson's theory predicts that the gas flowing out of the top of the bubble will be dragged back by moving bed particles and forced to re-enter the bubble at its base (Fig. 1.3d). Thus bubble gas circulates within a region of radius R_c, the region outside the bubble boundary being known as the "cloud"; the higher the bubble velocity the thinner becomes the cloud (Fig. 1.3e, f). The Davidson theory has provided much insight into bubble flow mechanisms in fluidized beds and despite its simplifying assumptions of purely circular or spherical bubbles it is still usefully applied in many circumstances.

Davidson's two-phase model was subsequently developed to take account of chemical reaction in a bubbling bed (Davidson and Harrison 1963). The model assumed a catalytic reaction to occur exclusively in the emulsion phase with first-order kinetics, the gas in this phase being either in plug flow or completely mixed. Interphase gas transfer was allowed for by a combination of molecular diffusion and gas throughflow and the bubble size was assumed to be constant throughout the bed. The model was later tested experimentally by Chavarie and Grace (1975) and found to be wanting in several respects notably the high value assumed for the interphase mass transfer rate resulted in reactant concentrations in the bubble and emulsion phases being predicted to be closely similar which was contrary to observation. Nevertheless despite its shortcomings Davidson's model formed the basis of a number of other more realistic approaches developed in this period (Kunii and Levenspiel 1968; Partridge and Rowe 1966; Kato and Wen 1969) although "there has not been any single model universally applicable to all processes carried out in (catalytic) reactors" (Ho 2003).

A second influential theoretical paper from this period was published by Jackson (1963) which purported to show that gas-solid fluidized beds are intrinsically unstable and that small disturbances will grow rapidly into bubbles under all circumstances. The work was unable to account for the stable, uniform expansion regime observed experimentally in beds of Group A powders between U_{mf} and U_{mb}. A number of authors subsequently attempted to explain the observed effect as being due to the formation of loose structures within the expanded regime, the structures being stabilised by inter-particle forces such as van der Waals or capillary condensation forces (Rietema 1973). The question of the stability of fluidized beds in the region of uniform expansion has been the subject of considerable controversy ever since and will be discussed in more detail in the context of the 1980s below.

1.2.3.1 New Processes

Propylene ammoxidation
This decade saw the commercialisation by the Sohio company of the catalysed reaction between propylene, ammonia and air (oxygen) to give acrylonitrile:

$$CH_3CH{=}CH_2 + NH_3 + 3/2\,O_2 = CH_2{=}CHCN + 3H_2O;\ \Delta H = -515\,\mathrm{kJ/mol}$$

Fixed-bed reactors have been used for the ammoxidation process but fluidized-bed systems are generally preferred for the following reasons:

- the highly exothermic reaction requires very efficient cooling to maintain bed temperatures in the required range of 400–510 °C and this is provided by cooling coils immersed in the bed which take advantage of the excellent bed-to-immersed surface heat transfer properties of fluidized beds.
- both reactants and products are flammable with air within wide limits but because of the flame-quenching action of the moving particles in a fluidized bed the three reactants can be admitted to the reactor without hazard.
- the good temperature control characteristic of fluidized beds enables the activity and life of the solid catalyst to be maintained for long periods without the need for regeneration.

The process will be discussed in more detail in Chap. 2.

Olefin polymerisation
A fluidized-bed process for the production of high-density polyethylene (HDPE) was introduced by Union Carbide (now Dow Chemical) in 1968. This so-called Unipol process has been extended in later years to the production of linear low-density polyethylene (LLDPE 1977), polypropylene (1985) and ethylene-propylene rubber 1997. The use of fluid-bed reactors in the Unipol and other similar processes developed by other companies confers a number of advantages over the alternative liquid-slurry reactors in use for polyolefin production since the 1930s, for example:

- efficient removal of the high exothermic heat of reaction
- good mass transfer between reactant gases and catalyst particles
- ease of downstream processing of the solid product and unreacted gases

These and other advantageous characteristics have led fluid-bed processes to be the dominant technology in the polyolefin field and will be discussed further in Chap. 2.

Riser cracking
The catalyst employed in the original FCC process (Sect. 1.2.1.1) was a silica-alumina material produced by mixing hydrogels of silica and alumina under carefully controlled conditions of pH, temperature and concentration. The resulting gel is filtered, washed, slurried with water and spray-dried to give a finely-divided amorphous material with a surface area of the order of 500 m^2/g. These were

gradually replaced in the 1960s by synthetic crystalline zeolites mainly based on the faujasite structure. The cracking activity of the crystalline zeolites is several orders of magnitude higher than that of the amorphous silica-aluminas but for practical purposes it is necessary to reduce this activity by incorporating small particles of zeolite into a silica-alumina base, a typical modern cracking catalyst containing from 3 to 25 weight percent zeolite. The use of these high-activity materials led to a radical re-design of cracking reactors, the bubbling-turbulent reactor section of the earlier designs such as the Standard Oil Model 4 being replaced by a vertical riser in which the cracking reaction takes place as the oil vapour and regenerated catalyst flow concurrently upwards. Risers can be up to 45 m in height.

1.2.4 The1970s

Geldart's breakthrough classification of powders into four groups designated A, B, C and D (Fig. 1.1) appeared (Geldart 1973) and was responsible for transforming the understanding of the behaviour of gas-fluidized beds. The boundary between Group A and Group B behaviour is defined as the line on the particle diameter-density plot for which $U_{mf}/U_{mb} = 1$. Then for Group A powders $U_{mf}/U_{mb} < 1$ and for Group B $U_{mf}/U_{mb} > 1$. From measurements on a wide range of particle-density combinations and air at ambient conditions Geldart was able to determine the boundary lines between the four groups as shown in Fig. 1.1.

Geldart's empirical classification has proved extremely useful in predicting the behaviour of various powders at and near the point of minimum fluidization but a theoretical explanation for the different types of behaviour did not emerge until some ten years later (Foscolo and Gibilaro 1984; see below).

During this period there was a significant increase in the number of publications relating to fast and circulating beds largely following pioneering work by Yerushalmi et al. at City College, New York (Yerushalmi et al. 1978) and by Reh at the Lurgi company in Germany (Reh 1971). There was also a worldwide growth of interest in fluidized-bed combustion and gasification (Yates 1983) and although this subsequently declined in some countries, notably the UK and USA, it remained a significant industrial activity in other areas such as China.

In 1970 the Mitsubishi Chemical Industries Company started up a commercial fluid-bed reactor of 18,000 tons/yr capacity for the production of maleic anhydride from a mixture of butadiene and butene (Kunii and Levenspiel 1991):

$$C_4H_8,\ C_4H_6 + O_2 = C_4H_2O_3;\ \Delta H = -1420\,\mathrm{kJ/mol}$$

The reactor operated in the turbulent-bed mode and was packed with cooling tubes to remove the heat of the highly exothermic reaction. A similar process but with a different reactor configuration was developed some years later by Lummus, BP and Du Pont. These processes will be discussed below.

1.2.5 The 1980s

A novel experimental technique introduced in this period was positron emission particle tracking, PEPT, (Parker et al. 1993) which enables the detailed movement of single particles of bed material to be followed. The principle of the technique is based on positron annihilation. A tracer particle labelled with a positron-emitting radionuclide is immersed in a fluidized bed and undergoes decay the positron-electron interaction resulting in the production of two γ-rays travelling in exactly opposite directions. Simultaneous detection of the two rays defines a line along which the annihilation occurred and several determinations over a short period coupled with triangulation enables the positional history of the tracer to be determined in three dimensions

Work on fluid-bed scaling laws (see Chap. 6) derived by matching dimensionless groups came into prominence with studies by Fitzgerald et al. (1984), Glicksman (1984) and Horio et al. (1986) while Grace (1986) published a useful extension of the Geldart powder classification based on dimensionless particle size and gas velocity.

The Particle-bed Model

This was the main theoretical advance in this decade and was developed initially by Foscolo and Gibilaro (1984) who on the basis of purely hydrodynamic considerations were able to account for the difference in behaviour of the four powder groups of the Geldart scheme. It will be recalled (Sect. 1.2.3) that Jackson's analysis of 1963 showed that gas-fluidized beds are inherently unstable and will form bubbles i.e. become "heterogeneous" at all gas velocities above U_{mf}; furthermore that other workers had proposed that the stability shown by Group A powders between U_{mf} and U_{mb} could be attributed to the formation of loose structures held together by inter-particle forces (Rietema 1973, 1991). Foscolo and Gibilaro took an alternative approach proposing that the homogeneous region of stability was purely the result of the hydrodynamic inter-action between particles and up-flowing gas and was quite independent of any form of inter-particle force. Although not apparent at the time of its first publication (Foscolo and Gibilaro 1984) their model was a direct consequence of Wallis's earlier analysis of two-phase flow (Wallis 1969) in which the stability of a two-phase system is determined by the relative values of the propagation velocities within the system of two waves, the dynamic or elastic wave and the kinematic or continuity wave both of which are functions of the physical properties of the system i.e. particle size and density, gas velocity, density, viscosity etc. Wallis's criterion had been applied to fluidized beds by Verloop and Heertjes (1970) who showed that the transition from homogeneous to heterogeneous fluidization could be described by the occurrence of shock waves which arose when the propagation velocity of the kinematic wave, u_K, exceeds that of the dynamic wave, u_D. Foscolo and Gibilaro extended this work and derived explicit expressions for u_K and u_D as follows (Gibilaro 2001).

$$u_D = \left[3.2gd_p(1-\varepsilon)\left(\rho_p - \rho_f\right)/\rho_p\right]^{0.5} \tag{1.8}$$

$$u_K = nu_t(1-\varepsilon)\varepsilon^{n-1} \tag{1.9}$$

In Eq. 1.9 the particle terminal-fall velocity, u_t, is found from:

$$u_t = \left[-3.809 + \left(3.809^2 + 1.832Ar^{0.5}\right)\right]^2 \mu_f/\left(\rho_f d_p\right) \tag{1.10}$$

where the Archimedes number, Ar, is calculated from the physical properties if the particles and fluid as:

$$Ar = gd_p^3\rho_f\left(\rho_p - \rho_f\right)/\mu_f^2 \tag{1.11}$$

The Richardson-Zaki exponent, n, is found from:

$$n = \frac{4.8 + 1.032\,Ar^{0.57}}{1 + 0.043\,Ar^{0.57}} \tag{1.12}$$

Then setting the minimum fluidization voidage, ε equal to 0.4 u_D and u_K may be calculated. The theory predicts stable fluidization i.e. uniform bed expansion if:

$$u_D > u_K \tag{1.13}$$

whereas if:

$$u_K > u_D \tag{1.14}$$

the bed will be unstable and will form bubbles at u_{mf}.

Two examples are illustrated in Fig. 1.4 from which it is clear that the bed of 150 μm particles will bubble at the point of minimum fluidization while that of the 70 μm material will expand uniformly up to a voidage of about 0.5 before forming bubbles. The relationship between model predictions and the Geldart plot are shown in Fig. 1.5.

A number of authors have questioned the validity of the particle-bed model on various grounds. Thus Batchelor (1988) and Jackson (2000) maintain that allowance must be made in the model formulation for momentum transfer due to particle collisions while Rietema et al. (1993) question the validity of the derivation used for the dynamic wave velocity. Thus the reasons for the stability or otherwise of fluidized beds is still open to question but nonetheless the agreement between the predictions of the particle-bed model and experimental observations of bed stability (Jacob and Weimer 1987; Foscolo and Gibilaro 1987) has provided a powerful justification for the assumptions on which the model is based (Valverde Millan 2012).

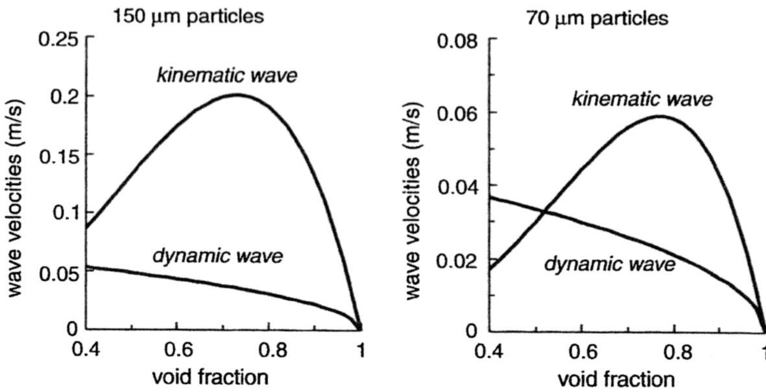

Fig. 1.4 Dynamic and kinematic wave velocities as functions of void fraction for the fluidization of alumina particles by air (ρ_p = 1000 kg/m^3; d_p = 150 and 70 μm; Gibilaro 2001)

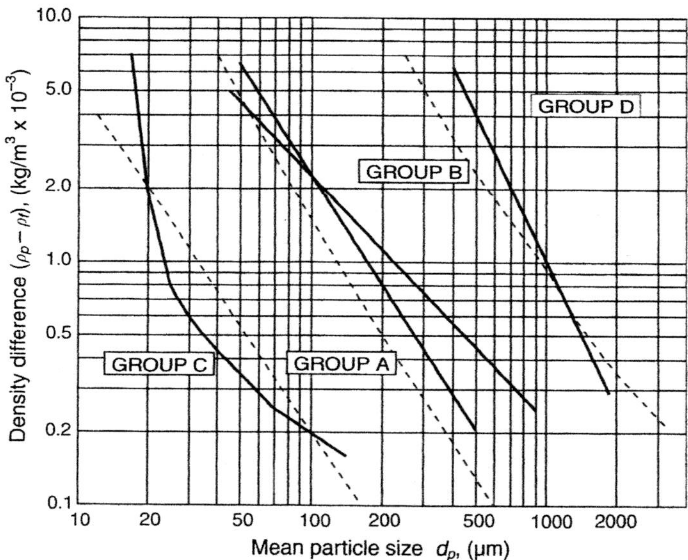

Fig. 1.5 Boundary predictions of the particle-bed model: heavy lines empirical boundaries due to Geldart (1973); broken lines model predictions (Gibilaro 2001)

Downer reactors

One of the disadvantages of the riser reactors referred to in Sect. 1.2.3.1 is that the core-annular structure of their flow pattern results in axial back-mixing of the solid particles giving them a broad residence time distribution. This has been found to cause over-cracking in some FCC units resulting in lower yields of gasoline and higher yields of coke and light gases. Although down-flow reactors were reported to have been used in Germany and the USSR in the 1960s and 1970s for the plasma ultrapyrolosis of coal (Zhu and Wei 1995) it was not until the 1980s that reports began to appear of their use in the process industries. Gartside (1989) described the development by the Stone and Webster company of a so-called "Quick Contact" or QC reactor in which reactant gas and solids flow co-current downwards in near plug flow. Earlier both Texaco and Mobil had patented downer reactors for the FCC process. Zhu and Wei (1995) list the advantages of downer systems for fast reactions with the intermediates as products;

(i) radial distributions of gas and solids are much more uniform than in riser reactors
(ii) axial dispersion of both gas and solids are two orders of magnitude less than in risers indicating near ideal plug flow conditions.

A state-of-the-art review of downer reactors was given by Jin et al. (2002).

1.2.6 The 1990s

A major development in this decade was the introduction via computational fluid dynamics (CFD) of numerical simulations of fluid bed hydrodynamics. Two broad classes of model have been studied the so-called Lagrangian and the Eulierian. In the former Newton's laws of motion are solved for individual particles and particle collisions are considered to be inelastic through friction and kinetic energy losses. Tsuji et al. (1993) were among the first to apply these discrete particle methods to fluidized beds. They used the distinct element method (DEM) developed by Cundall and Strack (1979) in which the inter-particle contact forces are calculated on the basis of simple mechanical models such as springs, dashpots and friction sliders. Lagrangian-Eulierian modelling of fluidized beds (Thornton and Kafui 2004) employs a DEM-based direct numerical integration of the individual particle trajectories which is then coupled to a continuum integration of the Navier-Stokes equations of fluid motion through an interphase interaction term. Discrete particle methods are valuable and give predictions in line with what is known from experiment; they are however of limited application at the present time owing to the extreme length of computer time necessary particularly in the case of small- particle systems.

The Eulerian approach is based on the "two-fluid" model originally proposed by Jackson (1971) which considers a fluidized bed to be made up of two inter-penetrating continua the fluidizing fluid and the particulate phase which is suspended in the up-flowing fluid. Granular kinetic theory (Gidaspow 1994) is used to determine the momentum balance of the solid phase while the corresponding balance for the gas phase is found by solving numerically the Navier-Stokes equations which are modified to include an inter-phase momentum transfer term. In order to explore the expected behaviour of a system it is then necessary to close the equations of motion by incorporating empirical or constitutive relationships between the variables which relate the acting forces and stresses to the local mean voidage, velocity and pressure fields.

1.2.7 The New Millennium: The First Decade

This period was notable for the increasing sophistication of experimental methods for studying fluidized beds, methods such as capacitance tomography, laser Doppler velocimetry, computerised tomography and acoustic sensing. These and other novel methods have been comprehensively reviewed by Horio et al. (2003). Subsequently, and building on the work of Grace and Baeyens (1986), Cheremisinoff (1986), Yates and Simons (1994) and Louge (1997), reviewed methods for measuring gas-solids distributions in fluidized beds including electrical capacitance tomography, X- and gamma-ray tomography, optical probes, capacitance probes and pressure transducers. Somewhat earlier Werther (1999) had given a detailed survey of methods used on an industrial scale to determine solids volume concentrations, solids velocities and mass fluxes and the vertical and horizontal distribution of gases and solids. Examples were suction probes, heat–transfer probes and capacitance probes.

Fluidized-bed combustion is currently witnessing a resurgence of interest particularly from the point of view of clean energy generation. Johnsson (2007) has reviewed recent developments and identified two main areas of application: (a) small and medium scale boilers of up to 100 MWth for heat only or combined heat and power generation, and (b) larger coal-fired boilers of up to 1000 MWth. These developments are reviewed below in Chap. 3.

Another development gaining prominence in this period and aimed at new processes for clean energy production is chemical looping combustion. Here a fuel is burnt in contact not with air but a solid oxygen carrier such as a metal oxide leading to a product gas stream high in concentration of carbon dioxide suitable for subsequent capture and sequestration; the reduced oxygen carrier is then regenerated with air. The work will be described below but a good summary of the state of the art at the end of the decade is given by Fan (2010).

During this decade there has been a growing interest in the use of ultrafine particles in the nanometre size range for the manufacture of materials such as semiconductors, drugs, cosmetics, foods, plastics, catalysts, paints, sunscreens and biomaterials (Ammendola and Chirone 2007). A number of research groups have explored the behaviour of nanoparticles under fluidization conditions and it was clear from the outset that, unlike in the case of particles in Geldart's Groups A and B in the absence of extraneous forces such as sintering or wetting, interparticle forces are a dominant feature. Strong cohesive forces between primary particles result in the formation of aggregates and it is these larger structures that determine the fluidization behaviour of the materials. The aggregates themselves can in many cases cohere into structures with such high strengths that fluidization in the normal way is not possible (Geldart Group C behaviour) but fluidization can be induced by the application of external forces such as mechanical vibration or sound waves (Hakim et al. 2005). Detailed consideration of the fluidization of nanoparticle systems is outside the scope of the present volume but the field has recently been comprehensively reviewed by Valverde Millan (2012).

1.2.8 Conclusions

In their Preface to the eleventh international conference on fluidization held in Naples in 2004 (Fluidization XI) the editors stated: "Fluidization is a strongly interdisciplinary area wherein process and chemical engineering, mechanical engineering, physics and mathematics still find, after more than fifty years of extensive research work, challenging problems to be dealt with" (Arena et al. 2004). The conference itself dealt with a wide range of topics from hydrodynamics in bubbling and circulating beds through to agglomeration, mixing and segregation, heat and mass transfer, dynamics and chaos, chemical, petrochemical,energy and environmental aspects. These and other areas have continued to attract academic and applied research but despite major advances in some areas much remains to be done before fluidized-bed reactors can be designed and scaled up with total confidence. The current state of the art constitutes the remainder of this book.

References

Ammendola P, Chirone R (2007) The role of sound vibration during aeration of nano-sized powders. Fluidization XII p 361–368, Engineering Conferences International, New York

Arena U, Chirone R, Miccio M, Salatino P (2004), Fluizization XI p xv, Engineering Conferences International, New York

Batchelor GK (1988) A new theory on the stability of a uniform fluidized bed. J Fluid Mech 193:75–110

Cheremisinoff A (1986) Review of experimental methods for studying the hydrodynamics of gas-fluidized beds. Ind Eng Chem Proc Des Dev 25:329–351

Chavarie C, Grace JR (1975) Performance analysis of a fluidized bed reactor II. Observed reactor behaviour compared with simple two-phase models. Ind Eng Chem Fundams 14:79–86

Cundal PA, Strack ODL (1979) A discrete numerical model for granular assemblies. Geotechnique 29:47–65

Davidson JF (1961) Symposium on fluidization; discussion. Trans Inst Chem Eng 39:230–232

Davidson JF, Harrison D (1963) Fluidised Particles. Cambridge University Press, Cambridge

Enos JL (1962) Petroleum progress and profits: a history of process innovation. MIT Press, Cambridge, MA

Fan LS (1989) Gas-liquid-solid fluidization engineering. Butterworth, Boston

Fan LS, Yang G (2003) Gas-liquid-solid three-phase fluidization, handbook of fluidization and fluid-particle systems, Yang W-C (ed) pp 765–809, Marcel Dekker, New York

Fan LS (2010) Chemical looping systems for fossil energy conversions. Wiley-AIChE, New York

Fitzgerald TJ, Bushnell D, Crane S, Shieh Y (1984) Testing of cold scaled bed modelling for fluidized-bed combustors. Powder Technol 38:107–120

Foscolo PU, Gibilaro LG (1984) A fully predictive criterion for the transition between particulate and aggregate fluidization. Chem Eng Sci 39:1667–1675

Foscolo PU Gibilaro LG (1987) Dynamic stability of fluidized suspensions: the particle-bed model. Chem Eng Sci 42:1489–1500

Gartside RJ (1989) QC—a new reaction system. Fluidization VI, Grace JR, Shemelt LW, Bergounou MA (eds) pp 25–32, Engineering Foundation, New York

Geldart D (1967) The fluidized bed as a chemical reactor: a review of the first 25 years. Chem & Ind 1474–1481

Geldart D (1973) Types of gas fluidization. Powder Tecnol 7:285–292
Gibilaro JG (2001) Fluidization-dynamics. Butterworth Heinemann, Oxford
Gidaspow D (1994) Multiphase flow and fluidization. Academic Press, London
Glicksman LR (1984) Scaling relationships for fluidized beds. Chem Eng Sci 39:1373
Grace JR, Clift R (1974) On the two-phase theory of fluidization. Chem Eng Sci 29:327–334
Grace JR (1986) Contacting modes and behaviour classification of gas-solid and other two-phase suspensions. Can J Chem Eng 64:353–363
Grace JR, Bi H (1997). Introduction to circulating fluidized beds. Chapter 1 in Circulating Fluidized Beds. Grace JR, Avidan AA, Knowlton TM (Eds). Blackie A&P, London
Grohse EW (1955) Analysis of gas fluidized solid systems by X-ray absorption. AIChE J 1:358–365
Hakim LF, Portman JL, Casper MD, Weimer AW (2005) Aggregation behaviour of nanoparticles in fluidized beds. Powder Technol 160:149–160
Ho TC (2003) Fluidization and Fluid-Particle Systems, Yang W-C (Ed) Chapter 9, Marcel Dekker, New York
Horio M, Nonaka A, Sawa Y, Muchi I (1986) A new similarity rule for fluidized bed scale-up. AIChE J 32:1466–1482
Horio M, Kobylecki R P, Tsukoda M (2003) Handbook of Fluidization and Fluid-Particle Systems, Yang W-C (Ed) Chapter 25, Marcel Dekker, New York
Jacob KV, Weimer AW (1987) High temperature particulate expansion and minimum bubbling of fine carbon powders. AIChE J 33:1698–1706
Jackson R (1963) The mechanics of fluidized beds: Part 1: The stability of the state of uniform fluidization. Trans Inst Chem Eng 41:13
Jackson R (1971) Fluidization, Davidson J F, Harrison D (Eds) Chapter 3, Academic Press, London
Jackson R (2000) The dynamics of fluidized particles. Cambridge University Press, Cambridge
Jahnig CE, Campbell DL, Martin HZ (1980) Grace JR Matsen JM (eds) Fluidization. Plenum Press, New York
Jin Y, Zheng Y, Wei F (2002) State of the art review of downer reactors. In: Grace JR, Zhu JZ, de Lasa H (eds) Proceeding 7th international conference on circulating fluidized beds, pp 40–60
Johnsson F (2007) Fluidized-bed combustion for clean energy. In: Bi X, Berruti F, Pugsley T (eds) Fluidization XII. Engineering Conferences International, New York, pp 47–62
Kato K, Wen CY (1969) Bubble assemblage model for fluidized bed catalytic reactor. Chem Eng Sci 24:1351–1369
Kunii D, Levenspiel O (1968) Bubbling bed model for flow of gas through a fluidized bed. Ind Eng Chem Fundams 7:446–452
Kunii D, Levenspiel O (1991) Fluidization engineering, 2nd edn. Butterworth-Heinemann, Boston
Lewis WK, Gilliland ER, Glass W (1959) Solid-catalysed reaction in a fluidized bed. AIChE J 5:419
Louge M (1997) Measurement techniques. In: Grace JR, Avidan AA, Knowlton TM (eds) Circulating fluidized beds. Blackie A&P, London
Mathis JF, Watson CC (1956) Effect of fluidization on catalytic cumene dealkylation. AIChE J 2:518
Morse RD, Ballou CO (1951) The uniformity of fluidization, its measurement and use. Chem En Prog 47:199–211
May WG (1959) A model for chemical reaction in a fluidized bed. Chem Eng Prog 55(12):49
Mickley HS, Fairbanks DF (1955) Mechanism of heat transfer in fluidized beds. AIChE J 1:374
Parker DJ, Hawkesworth MR, Beynon TD, Bridgwater J (1993) Process engineering studies using positron-based imaging techniques. In: Beck MS (ed) Tomographic techniques for process design and operation. Computational Mechanics Publications, Southampton, UK
Partridge BA, Rowe PN (1966) Chemical reaction in a bubbling fluidized bed. Trans Inst Chem Eng 44:335–348
Reh L (1971) Fluidized bed processing. Chem Eng Prog 67(2):58–63
Richardson JF, Zaki WN (1954) Sedimentation and fluidization. Trans Inst Chem Eng 32:35

Rietema K (1973) The effect of interparticle forces on the expansion of a homogeneous gas-fluidized bed. Chem Eng Sci 28:1493–1497

Rietema K, Cottar EJE, Piepers HW (1993) The effects of interparticle forces on the stability of gas-fluidized beds. Chem Eng Sci 48:1687–1697

Rietema K (1991) The dynamics of fine powders. Elsevier, London

Rowe PN (1964) Gas bubbles in a two-dimensional fluidized bed. Chem Eng Prog 60(3):75

Shen CY, Johnstone HF (1955) Gas-solid contact in fluidized beds. AIChE J1:349–354

Tsuji Y, Kawaguchi T, Tanaka T (1993) Discrete particle simulation of two-dimensional fluidized bed. Powder Technol 77:79–87

Thornton C, Kafui KD (2004) Fully 3D DEM simulations of fluidized beds including granulation. Arena U, Chirone R, Miccio M, Salatino P (eds) Fluidization XI. Engineering Conferences International, New York, p 195–202

Toomey RD, Johnstone HF (1952) Gaseous fluidization of solid particles. Chem Eng Prog 48:220–236

Valverde Millan JM (2012) Fluidization of fine powders. Springer

Verloop J, Heertjes PM (1970) Shock waves as a criterion for the transition from homogeneous to heterogeneous fluidization. Chem Eng Sci 25:825–832

Wallis GB (1969) One dimensional two-phase flow. McGraw Hill, New York

Werther J, Molerus O (1973) the local structure of gas fluidized beds 2: The spatial distribution of bubbles. Int J Multiphase Flow 1:123–138

Werther J (1999) Measurement techniques in fluidized beds. Powder Technol 102:15–36

Yagi S, Kunii D (1955) In: Proceeding of 5th International Symposium on Combustion 231

Yasui G, Johnanson LN (1958) Characteristics of gas pockets in fluidized beds. AIChE J 4:445–452

Yates JG (1983) Fundamentals of Fluidized-bed chemical processes. Butterworths, London

Yates JG, Simons SJR (1994) Experimental methods in fluidization research. Int J Multiphase Flow 20(suppl):297–330

Yerushalmi J, Cankurt NT, Geldart D, Liss B (1978) Flow regimes in vertical gas-sold contact systems. AIChE Symp Ser 174(176):1–12

Zhu JX, Wei F (1995) Recent developments in downer reactors and other types of short contact time reactors. Large J-F, Laguerie C (eds) Fluidization VIII, Engineering Foundation, New York, p 501–510

Chapter 2
Catalytic Processes

Abstract The chapter begins with a brief analysis of the advantages and disadvantages of fluidized-bed reactors compared with alternatives such as fixed beds for solid-catalysed gas-phase processes coupled with some general points about reactor operation. The points made are emphasised in the descriptions that follow of the most prominent technologies currently employing gas-solid catalytic reactions. Olefin polymerization is traced from its introduction in the 1960's to modern-day variants employing condensing-mode operation. Operational problems associated with electrostatic charging of the fluidized polymer particles are discussed. Processes for the oxidation of n-butane to maleic anhydride are presented, particular attention being paid to the DuPont circulating fluidized-bed process which although ultimately ending in failure demonstrated important aspects of plant design and operation. Well-established processes for the ammoxidation of propylene to acrylonitrile are discussed the emphasis being on that developed by Sohio. Processes for the production of vinyl chloride monomer and vinyl acetate monomer are described briefly. A section on gas-to-liquid technologies describes the classic Synthol process as well as more recent developments converting synthesis gas to methanol and thence to gasoline and light olefins. The chapter concludes with a consideration of fluidized catalytic cracking, arguably the most important catalytic reaction in all industry. In each case the emphasis is on process chemistry, catalyst formulation, reactor configuration and operation and reactor/process modelling.

2.1 Introduction

Processes based on the use of heterogeneous catalysts lie at the heart of the chemical and allied industries and of the many types of reactor employed in these processes the fluidized bed offers a number of advantages over alternative designs. In beds operated at high gas velocities in the bubbling and turbulent regimes referred to earlier bed particles are in constant motion and as a result are well mixed. Thus any hot spots that may arise in an exothermic reaction are rapidly quenched and the bed operates essentially isothermally. A further consequence of

J.G. Yates and P. Lettieri, *Fluidized-Bed Reactors: Processes and Operating Conditions*, Particle Technology Series 26,
DOI 10.1007/978-3-319-39593-7_2

good solids mixing is that heat transfer between bed material and immersed surfaces is highly efficient so that with appropriate design of heat exchangers beds may be operated within closely controlled temperature limits. Another advantage is the ability to introduce multiple feeds directly into the bed without the necessity of pre-mixing, something difficult to achieve with a fixed-bed reactor. Further the liquid-like properties of fluidized solids enable them to be transferred smoothly from one reactor to another for catalyst regeneration and reheating as in the FCC process. Once the catalyst particles are fluidized the pressure drop across the reactor, unlike for fixed-bed reactors, remains more or less constant with increasing gas velocity so for high gas throughputs and equivalent bed heights pressure drops through fluidized beds are lower than through fixed beds leading to lower compression costs in the former case. Another advantage is that the use of small particles ensures effectiveness factors are close to unity (Jiang et al. 2003).

There are however a number of disadvantages of fluid-bed operation such as gas back-mixing leading to the formation of unwanted side products, particle attrition leading to loss of catalyst through cyclones, and erosion of bed internals by the sand-blasting effect of the moving particles. Because of the frequently complex hydrodynamics of fluidized beds scale-up can be challenging. These and other considerations will be explored in the context of the individual processes described below.

The majority of fluidized-bed reactors used in catalytic processes are operated at high gas velocities in the turbulent, circulating or fast regimes (Fig. 1.1).

The reactors have several features in common—cylindrical vessels of relatively high aspect ratio fitted at their lower end with a gas distributor at their upper end with one or more cyclone separators and often packed in between with heat-transfer tubes or coils. The distributor or grid spans the cross-section of the bed at its base above the windbox or plenum chamber into which the fluidizing gas is admitted. Distributors should be designed so as to maintain the bed solids in uniform motion and prevent the formation of defluidized zones, operate for long periods without plugging, minimise attrition of bed particles and minimise the leakage of particles into the plenum. A number of different designs have been used and classified according to the direction of gas entry: upward, laterally or downward (Karri and Werther 2003) through bubble caps, nozzles, spargers, conical grids and pierced sheets. The entry points are frequently arranged on a square or triangular pitch, open holes often being fitted with shrouds, short lengths of pipe centred over the holes, to help reduce particle attrition. In order to maintain a constant and uniform fluidization in the bed it is necessary to maintain a sufficiently high pressure drop, Δp_D, across the grid. This is normally considered in relation to the pressure drop across the bed as a whole, Δp_B, the value depending on the direction of gas entry. Thus Karrie and Werther (2003) propose:

$$\Delta p_D = 0.3 \Delta p_B$$

for upward and laterally directed flow, and:

$$\Delta p_D = 0.1\Delta p_B$$

for downward flow and furthermore that the pressure drop across a large-scale grid should never be less than 2500 Pa.

The function of the cyclone at the exit of the bed is to separate gases from solids so as to minimise particulate emissions and return catalyst material to the bed. They act by causing a centrifugal force to be acted on the particles forcing them to the wall of the vessel where they lose momentum and spiral downwards via a dipleg and flapper valve back to the bed. The separated gas flows upwards and out of the unit.

Cyclones have the advantages of having no moving parts, being inexpensive to construct, having low pressure drops and low maintenance costs. To increase solids collection efficiencies they are sometimes operated in series, two-stage cyclones in fluidized catalytic cracking regenerators, for example, having efficiencies of over 99.999 % (Knowlton 2003).

Other detailed aspects of reactor design and operation will be discussed in the context of the individual processes that follows.

2.2 Some Individual Processes

2.2.1 Olefin Polymerization

2.2.1.1 Process Background

As mentioned above (Sect. 1.2.3.1) the Union Carbide company introduced the first fluid-bed process for the polymerization of olefins in 1968 and since then a number of other companies, notably Sumitomo, Mitsui, BASF, ExxonMobil and BP, have commercialized similar technology. A schematic of the Unipol process is shown in Fig. 2.1.

The reactant gases, ethylene and co-monomers such as butene, are fed along with very small silica-supported catalyst particles into the base of the reactor at three to six times the minimum fluidization velocity of the larger bed particles (Kunii and Levenspiel 1991). The reactor contains preformed polyethylene particles maintained at 40–120 °C and 10–40 kPa pressure. As the catalytic reaction proceeds the polymer particles grow up to 5000 μm in size the product being withdrawn from the reactor so as to keep the bed at more or less constant volume. They are removed at a point above and close to the distributor through sequential operation of a pair of timed valves and separated from unreacted gases which are then recycled. Temperature control is crucial to successful operation since the

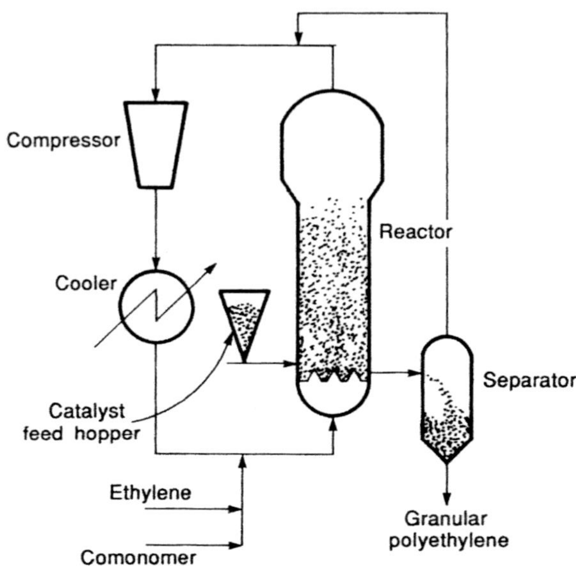

Fig. 2.1 The Unipol process for the production of polyethylene (Kunii and Levenspiel 1991)

reaction is highly exothermic and the bed operates close to the melting point of the polymer. Indeed the production rate is limited by the maximum rate at which heat can be removed from the reactor. As a result per-pass conversion is kept low at around 1–2 %, unreacted gases being recycled and cooled before being readmitted to the reactor. Because of the potential build-up of polymer deposits internal cooling coils cannot be used so early units such as that shown employed external heat exchangers. The Unipol process and similar variants have been comprehensively reviewed by Xie et al. (1994). Modern designs operate in the condensing mode where a proportion of the recycle gas is cooled to below its dew point to form liquid before its readmission to the reactor; the latent heat of evaporation of the liquid is then used to aid temperature control. Operation in the condensing mode has enabled newer plants to increase capacity by up to 200 % (Jazayeri 2003) i.e. of the order of 200,000 tonnes per year.

The beds operate in the turbulent flow regime. Bed particles have the physical properties of Group B materials (d_p = 300–5000 μm, ρ_p = 550–850 kg/m^3) but owing to the high pressure of operation they show Group A behaviour i.e. smooth fluidization and extensive emulsion-phase expansion (Burdett et al. 2001).

The process may be used to produce a wide range of products including homo- and copolymers of ethylene and propylene or copolymers containing one or more C_3 to C_8 alpha-olefins. Linear low density polyethylene (LLDPE) is made from copolymers of ethylene, butene, 4-methylpentene-1 or hexene while high density polyethylene (HDPE) results from homopolymerization of ethylene or copolymerization with butene, pentene, hexene, 4-methylpentene-1 or octene.

2.2.1.2 Condensing-Mode Operation

In a version of the process developed by BP (Chinh et al. 1998)the main body of the reactor is a cylindrical vessel with a gas distributor at its lower end, the distributor being a flat or dished plate perforated by equally distributed holes some 5 mm in diameter; distributor grids are often designed with relatively large holes without nozzles (Yamamoto et al. 1998). In the upper part of the reactor with an expanded cross-sectional area the gas velocity is decreased enabling some entrained particles to be returned to the bed below. The gases exiting the reactor pass to a cyclone where fines are separated, the gas then passing to a heat exchanger and a compressor; a second heat exchanger removes the heat of compression. The two exchangers are operated so as to cause a proportion of the recycle gas to be condensed, the gas-liquid recycle stream then passing to a separator where the liquid condensate is removed and pumped back into the reactor. The separated gas is recycled to the bed along with the amount of monomer/comonomer required to replace that consumed during the polymerization. In the BP process the liquid condensate is injected into the fluidized bed through a multi-orifice nozzle located at a distance above the gas distributor where the bed temperature has reached that required for the polymerization. The liquid condensate may be a condensable monomer such as butene, hexene or octene or an inert condensable liquid such as butane, pentane or hexane. The mass ratio of liquid to total gas injected can cover a wide range e.g. 6:100 to 25:100 depending on the activity of the catalyst and the required production rate. Hydrogen is added to the recycle stream to act as a chain-transfer agent in the polymerization reaction to control the molecular weight of the product.

2.2.1.3 Catalysts

Heterogeneous catalysts for the polymerization of alpha-olefins were introduced by Ziegler and Natta in the 1950's. They were based on transition metal halides such as $TiCl_4$ in combination with an organo-aluminium compound such as aluminium tri-ethyl. Depending on the exact formulation of the catalyst and the processing conditions the polymer products were found to have varying stereospecific structures at the resulting chiral centres. Thus polypropylene could have one of three so-called tacticities: isotactic where the chiral centres were all similarly orientated i.e. all either *d* or *l*, syndiotactic where they were alternately *d* and *l*, or atactic where the chiral centres were randomly distributed. Similar structures were found to result from using chromium-based coordination compounds developed around the same time by the Phillips company.

In recent years a new class of polymerization catalysts has been introduced constituting a major advance on the earlier materials. These are the metallocenes one example of which is dicyclopentadieneylzirconium dichloride $(C_5H_5)_2ZrCl_2$ activated in a similar manner to Ziegler-Natta catalysts by an aluminium compound methylaluminoxane (MAO) $Al_4O_3(CH_3)_6$. The metallocenes are soluble in hydrocarbons and with easily variable structures enable the properties of the resulting polymers to be accurately predicted with controllable molecular weight distributions and tacticities. In addition they are 10–100 times more active than Ziegler-Natta systems, a combination of zirconocene and MAO allowing the polymerization of 100 tonnes of ethylene per gram of zirconium (Kaminsky 1998); they are so active that there is no requirement to separate them from the polymer product at the end of the process.

To be used in fluidized-bed polymerization reactors the catalysts must be impregnated into solid supports such as particles of silica, alumina or preformed polyolefin although silica is the preferred support. The amounts of metallocene and activator are in the range 0.01–0.5 and 0.5–20 mmol per gram of carrier particles respectively (Yamamoto et al. 1998). The porous silica-based catalyst particles are normally produced by spray drying.

During reaction polymer growth occurs within catalyst pores until hydraulic pressure fractures the particle; an outer polymer shell then develops and the particles continue to fragment exposing additional active sites (Burdett et al. 2001). The process has been described in detail in Xie et al. (1994).

2.2.1.4 Electrostatic Effects

The polymer particles produced in the fluidized-bed polymerization process have dielectric properties and as a result of their frequent frictional contacts with other bed particles and with the walls of the reactor they generate electrostatic charges. This is known as "triboelectrification" and is frequently observed in fluidized beds of dielectric or refractory particles (Boland and Geldart 1971). In the case of olefin-polymerization reactors electrostatic forces can cause particles to adhere to the reactor walls especially on the sloped region of the upper disengaging zone where they overheat and fuse together due to the exothermic heat of reaction. Particles behaving in this way are referred to as "sheets" and they can grow up to several square metres in size and some centimetres thick (Hendrickson 2006). If the sheets fall to the base of the reactor they can block the holes of the distributor plate leading to maldistribution of fluidizing gas and ultimately to loss of fluidization altogether. This necessitates shut-down for clean-up and consequent loss of production. This clearly presents a major problem and much effort has been put into finding solutions (Burdett et al. 2001). A common method of reducing the build-up of electrostatic charges in fluidized beds is to humidify the fluidizing gas but owing to the poisoning effect of water on the organometallic catalysts this cannot be used in the present case.

Various techniques for dissipating electrostatic charges have been proposed by the operating companies and the patent literature contains many examples including the use of special wall coatings (Fulks et al. 1989) and static agents (Goode et al. 1989). The most effective technique however appears to be injecting antistatic agents such as quarternary ammonium salts into the bed to increase the surface conductivity of the particles (Fischer et al. 2000) although the problem does not appear to have been entirely eliminated in all cases (Moughrabiah et al. 2012).

2.2.1.5 Reactor Modelling

(i) Choi and Ray (1985) analysed the dynamics of a fluidized-bed polyethylene reactor using a two-phase (emulsion and bubble) model and showed them to be prone to unstable behaviour and temperature oscillations. Their work was later extended by McAuley et al. (1994) and further by McAuley et al. (1995) who based their studies on a simplified well-mixed model that showed close similarities with the earlier work. McAuley et al. (1995) first examined the behaviour of the reactor itself and showed the existence of three steady states, a lower at around 300 K which, depending on the catalyst feed rate, is either stable or unstable, a middle state at around 700 K that is always unstable, and an upper state at around 1200 K that is always stable. The only area of interest however is that below the melting point of polyethylene around 400 K. Adding a recycle stream and external cooler to the model showed major differences from the reactor-only case where for a range of catalyst feed rates no steady state was found to exist, limit cycle behaviour being obtained. The authors examined the effect of an ethylene partial-pressure controller and showed that higher partial-pressure set points led to runaway towards a higher-temperature steady state. Further, feed-back temperature control was always essential to maintain steady-state operation.

(ii) Fernandez and Lona (2001) developed a complex model of a polyethylene reactor based on a three-phase description of the system: gaseous bubble phase, gaseous emulsion phase and solids polymer particle phase. They assumed the bubble-phase and emulsion-phase gases to be in plug flow upward and the emulsion-phase solids to be in plug flow downward. Polymer particles were assumed to have a wide size distribution and to segregate according to their size and mass. Their kinetic model assumed the successive steps of catalyst formation, polymerization initiation, propagation, chain transfer, termination and catalyst deactivation, the final model being composed of 165 differential equations. Numerical solution required five sets of iterations; first the system was solved without accounting for the bubble phase in order to evaluate the total monomer consumption. Emulsion-gas and bubble concentrations were then determined followed by overall energy balances. The model showed good agreement with results reported in the literature and in patents. The work was later extended by Fernandez and Lona (2004).

(iii) Kiashemshaki et al. (2006) developed a two-phase model in which the fluidized bed was divided into four sections in series, the gas phase being in plug flow and the emulsion phase completely mixed while Ibrahema et al. (2009) explored a model consisting of four phases: bubbles, cloud, emulsion and solids. Shamiri et al. (2011) developed a novel two-phase model for propylene polymerization in a fluidized bed but at the time of publication no experimental validation had been obtained.

Despite the considerable body of published work in this area it remains questionable whether two- and three-phase models of the kind described above are capable of giving an adequate description of polyolefin reactors. The Unipol and other similar reactors are known to operate in the turbulent regime where gas bubbles are at best ill-defined and may well be absent altogether as definite entities. It may well be that generalised bubbling/turbulent models such as that proposed by Thompson et al. (1999) are to be preferred for these systems as may be the work of Alizadeh et al. (2004) whose tanks-in-series model of a polyethylene reactor is based on the turbulent model of Cui et al. (2000). Also noteworthy are the two dynamic models described by Secchi et al. (2013). The model of Thompson et al. (1999) treats the turbulent condition as being transitional between the purely two-phase bubbling and slugging regimes and the homogeneous single-phase structure of axially-dispersed plug flow, merging one into the other using the probabilistic averaging of key parameters that vary continuously with superficial gas velocity. In other words the model predicts a smooth transition from two-phase fluidization to single-phase axially-dispersed flow. The model was implemented in the gPROMS software from Process Systems Enterprises Ltd and shows great promise for application to general high-velocity turbulent-flow systems.

2.2.2 *n-butane Oxidation to Maleic Anhydride*

The main uses of maleic anhydride are for the production of unsaturated polyester resins (ca 41 %), butane diol (14 %), maleic copolymers (8 %) and tetrahydrofuran (7 %); installed world capacity is of the order of 1.2 Mt/a (Chiusoli and Maitlis 2008). As well as being an important industrial solvent tetrahydrofuran in turn is used in the production of the segmented polyurethane spandex and elastane fibres such as Lycra and copolyester elastomers such as Hytrel.

2.2.2.1 Process Background

Traditionally produced by the catalytic oxidation of benzene maleic anhydride is now made almost exclusively by the partial oxidation of n-butane over supported vanadium phosphorus oxide catalysts $(VO)_2P_2O_7$:

$$C_4H_{10} + 3.5O_2 = C_4H_2O_3 + 4H_2O; \Delta H = -1260\,\mathrm{kJ/mol}$$

However a complex network of reactions underlies this simple stoichiometry, butene, butadiene and furan having been suggested as intermediates while carboxylic acids and oxides of carbon are also formed in side reactions.

Both multi-tubular fixed-bed and fluidized-bed reactors are widely employed. The reaction is highly exothermic and temperature control is important to maintain catalyst activity and selectivity. In multi-tubular fixed-bed systems this is achieved by circulating a heat-transfer medium of molten salts, the hydrocarbon concentration being maintained below the flammability limit of 1.8 mol% n-butane in air. In fluidized beds temperature control is achieved by means of internal cooling coils and feed concentrations of up to 4 mol% n-butane in air are possible due to the flame-arresting properties of the bed particles (Contractor 1999). Axial solids mixing and the associated gas backmixing however result in a significant loss of selectivity. The recent development of a circulating fluidized-bed reactor (CFB) for n-butane oxidation avoids this problem by carrying out the hydrocarbon oxidation step and the catalyst re-oxidation step in two separate but connected sections of a looped system (see below).

2.2.2.2 The Catalyst

The catalyst used in the selective oxidation of n-butane to maleic anhydride is vanadyl pyrophosphate, $(VO)_2P_2O_7$, often referred to as vanadium phosphoprus oxide or VPO. Owing to its commercial importance a great deal has been published on this material in both the scientific and patent literature over the years (e.g. Blum and Nicholas 1982; Contractor et al. 1987; Bergna 1988; Centi 1993; Patience et al. 2007) and special methods have been developed for producing catalyst suitable for use in fluidized-bed reactors. The VPO precursor material is normally made from vanadium pentoxide and phosphoric acid in an organic medium and following filtration and drying is mixed with polysilicic acid as a slurry then spray dried to form microspheroidal particles. For fluidized-bed application the particles must be attrition resistant and this can be achieved by mixing the dried powder with colloidal silica prior to spray drying; however this can cause a significant loss in selectivity (Blum and Nicholas 1982).

Contractor et al. (1987) describe a method for producing an attrition-resistant catalyst suitable for use in the DuPont Circulating Fluidized-bed process. Thus vanadium pentoxide (100 g) is stirred into a mixture of isobutanol (1L) and benzyl alcohol (150 g) and refluxed for 12 h. 85 % phosphoric acid (150 g) is slowly added and again refluxed for 12 h. The precursor product, vanadyl hydrogen phosphate hemihydrate $VOHPO_4 \cdot 0.5H_2O$, is then cooled, filtered, dried and milled to give particles of 1–2 μm which are then slurried with freshly prepared polysilicic acid hydrogel to give a final solid composition containing 10 % silica. The slurry is then spray dried at 250 °C to give microspheres of mean diameter 70 μm with a strong porous layer of SiO_2 on the outer surface (Bergna 1988). In the final stage the microspheres are calcined and activated with air at 6 bar and 390 °C for four hours (Patience et al. 2007). The particles have a thin coating of silica that is durable but porous to reactants and products. The catalytic reaction of n-butane over vanadyl pyrophosphate is complex being a 14-electron oxidation involving the abstraction of 8 hydrogen atoms and the insertion of three oxygen atoms the processes occurring entirely on the adsorbed surface. Despite much research (Centi 1993) there is still uncertainty as to the nature of the active centre responsible for the catalysis but it is widely accepted that the process involves a redox reaction according to the Mars-van Krevelen mechanism (Mars and van Krevelen 1954) in which the V^{5+} oxidation state becomes reduced to V^{4+}. Industrial practice is to reoxidise the V^{4+} species on the catalyst surface with molecular oxygen either by co-feeding with the alkane or by regenerating in a separate unit. From limited data on industrial catalysts Centi (1993) concluded that the stable active catalyst is not fully achieved until the butane oxidation has been carried out for 200–500 h; such catalysts are said to be “equilibrated”.

Mills et al. (1999) studied the redox kinetics of $VOPO_4$ phases with n-butane and air and showed that the reduction stage leading to the formation of maleic anhydride could be described by a rate expression first order in n-butane and approximately one-third order in the concentration of lattice oxygen. They also showed that during reduction the lattice oxygen species on the catalyst surface react with n-butane thereby establishing a positive gradient for diffusion of sub-surface oxygen to the surface. In the regeneration stage the depleted surface oxygen sites are replaced by oxygen from the air. The kinetic parameters determined by Mills et al. (1999) were applied by Roy et al. (2000) in modelling the riser reactor of the DuPont CFB process (see below).

2.2.2.3 Mitsubishi Process

A commercial plant with a nameplate capacity of 18,000 t/a of maleic anhydride was put into operation in 1970. A schematic flow diagram of the process is shown in Fig. 2.2 (Contractor and Sleight 1987) and an outline of the reactor in Fig. 2.3 (Kunii and Levenspiel 1991).

The feed was the crude C–4 fraction from a naphtha cracker and the reacted gases were fed to a quench tower where the anhydride was absorbed in water to form an aqueous solution of maleic acid and distilled to give the final product. The reactor itself was 6 m in diameter and 16 m high and operated at 400–500 °C and 4 bar pressure. The bed was fluidized with air and the vaporized hydrocarbon feed introduced through a specially designed distributor with hundreds of nozzles (Kunii and Levenspiel 1991). The vessel contained vertical cooling coils generating high-pressure steam; the coils were also believed to restrict gas backmixing and hence to increase catalyst selectivity although work by DuPont proved this to be ineffective. The catalyst was a silica-supported VPO with a size range of 20–200 μm.

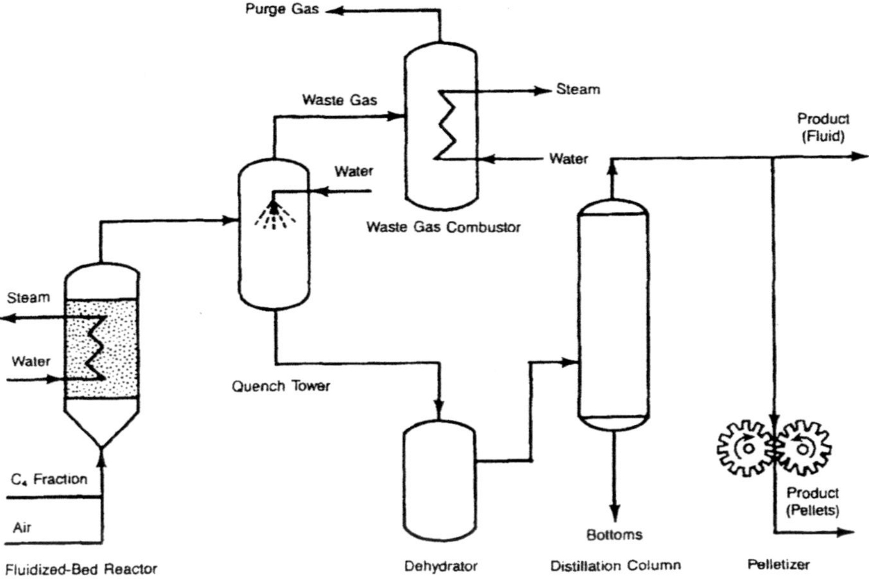

Fig. 2.2 Mitsubishi process for the production of maleic anhydride (Contractor and Sleight 1987)

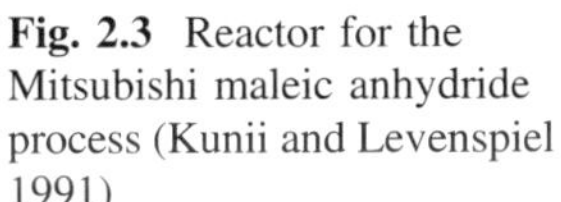

Fig. 2.3 Reactor for the Mitsubishi maleic anhydride process (Kunii and Levenspiel 1991)

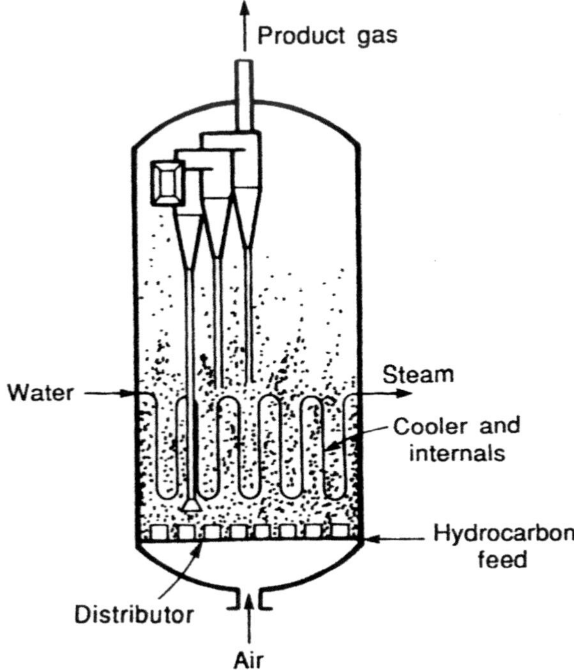

2.2.2.4 ALMA Process

ABB Lummus/Lonza under the brand name ALMA have developed a process using a bubbling bed configuration similar to that of Mitsibishi (Fig. 2.3) (Contractor and Sleight 1987; Dente et al. 2003). The reactor is 7 m in diameter and operates at bed temperatures of the order of 240 °C and pressures of 2–4 bar; the n-butane concentration is 4–5 % v/v. The n-butane conversion exceeds 80 % but molar yields of maleic anhydride are limited to 50–55 % (Dente et al. 2003).

The reactor performance is discussed further below.

2.2.2.5 DuPont Circulating Fluidized-Bed Process

The underlying principle of this process was to separate into two reactors the n-butane oxidation step and the reoxidation step of the reduced catalyst, analogous to the cracking-regeneration stages of the FCC process. The VPO catalyst thus acts as an oxygen carrier for the conversion. The process was developed by DuPont over a period of some 20 years from laboratory-scale riser reactors through a pilot plant to a full-scale commercial plant situated at Asturias in northern Spain. The history of the development has been documented in a number of publications (e.g. Contractor and Sleight 1987; Contractor 1999; Patience and Bockrath 2010).

The pilot plant, a scaled down version of the projected commercial unit, consisted of five vessels: 0.3 m × 6 m fast fluidized bed, 0.15 m × 24 m riser, 0.44 m diameter riser stripper, 0.53 m diameter catalyst regenerator, 0.44 m diameter regenerator stripper (Fig. 2.4).

The fast-bed section of the commercial reactor designed on the basis of the pilot-plant results is shown diagrammatically in Fig. 2.5. Recycle gas and fresh n-butane were fed into the bed through a grid-plate distributor at rates of between 0.43 and 0.93 m/s. Owing to the high oxygen demand of the reaction additional oxygen was introduced into the fast bed via three rows of spargers situated below two banks of cooling coils. Regenerated catalyst entered the bed from a slanted standpipe that changed from a circular geometry 1.2 m in diameter at the exit of the regenerator to an oval geometry at the inlet to the fast bed; the bed itself was 4.2 m in diameter and 11.5 m tall. The gas-solid suspension exited the fast bed into the 1.8 m × 28.5 m riser where it passed in plug flow to a rough-cut cyclone then to a stripper and thence to an external cyclone. The product gases passed to a water absorber where the anhydride product was converted to maleic acid a large fraction of the effluent gases being recycled to the fast bed. The maleic acid solution was then hydrogenated to give tetrahydrofuran the final product. On separation from the product gases the VPO catalyst was passed from the stripper to the regenerator, a 4.0 m × 16 m turbulent-flow bed fluidized with air and fitted with horizontal cooling coils. The catalyst inventory in the system was 175 tonnes and solid circulation rates of up to 7000 tonnes/h were achieved. At these high circulation rates with the catalyst flowing through the cyclone, between slide valves and cooling coils a certain degree of attrition was inevitable. Based on the pilot-plant

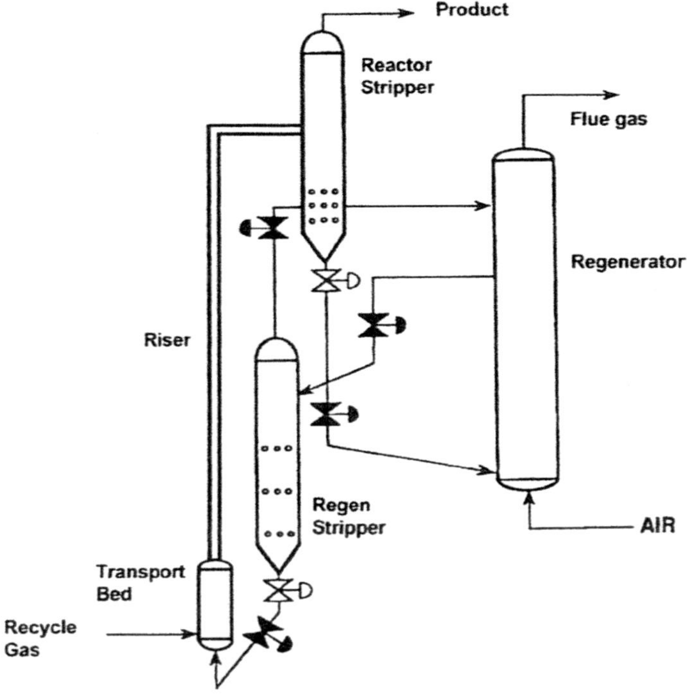

Fig. 2.4 Pilot-plant configuration for the DuPont CFB process (Patience and Bockrath 2010)

Fig. 2.5 Reactor for the DuPont CFB process (Patience and Bockrath 2010)

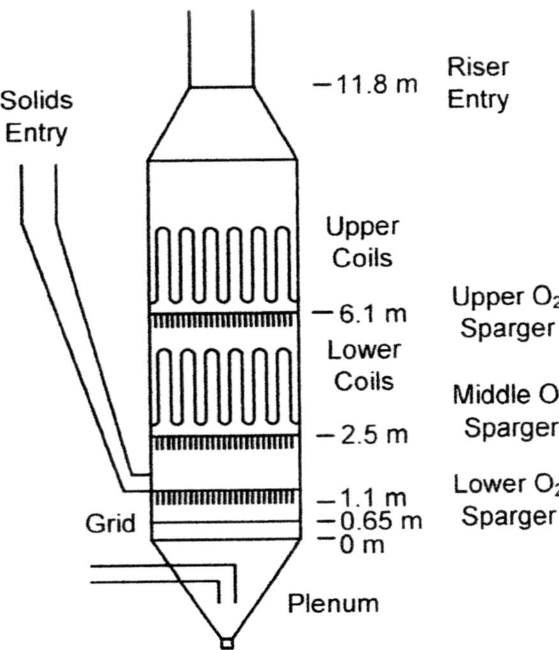

performance attrition rates of 5–15 kg/h were expected but lower values in the region of 1.0 kg/h were obtained in practice (Patience and Bockrath 2010).

Some of the operating problems that were never completely solved included incomplete regeneration of the catalyst which necessitated extra oxygen being fed to the fast bed, poor radial distribution of solids in the fast bed and backflow of gas from the fast bed to the regenerator standpipe. Nevertheless the plant operated for a decade after its start-up in 1996 before being finally shut down and dismantled.

2.2.2.6 Reactor Modelling

(i) Moustoufi et al. (2001) studied the performance of a fluidized-bed n-butane-oxidation reactor using three different models:

 (a) A simple two-phase model in which all the reaction takes place in the emulsion phase
 (b) A so-called "dynamic" two-phase model that considers some reaction to occur in the bubble phase as well as in the emulsion
 (c) A model in which the solids are assumed to be uniformly distributed in the bed with a constant voidage the gas passing through in plug flow.

The reaction kinetics used for the model predictions were based on the work of Centi et al. (1985) who determined the oxidation-rate parameters from isothermal steady-state fixed-bed reactor data and proposed the following triangular reaction network where MAN indicates maleic anhydride:

$$\begin{aligned} \text{n-C}_4\text{H}_{10} + 3.5\text{O}_2 &= \text{MAN} + 4\text{H}_2\text{O} \\ \text{MAN} + 4.5\text{O}_2 &= 4\text{CO}_2 + \text{H}_2\text{O} \\ \text{n-C}_4\text{H}_{10} + 6.5\text{O}_2 &= 4\text{CO}_2 + 5\text{H}_2\text{O} \end{aligned}$$

The corresponding rate equations are:

$$r_1 = r_{MAN} = \frac{k_1 K_B c_B c_O^{\alpha}}{1 + K_B c_B} \tag{2.1}$$

$$r_2 = r_{CO_2} = k_2 c_O^{\beta} \tag{2.2}$$

$$r_3 = -r_{MAN} = k_3 c_{MAN} \left(\frac{c_O^{\gamma}}{c_B^{\delta}} \right) \tag{2.3}$$

where r_1 is the rate of formation of maleic anhydride from n-butane, r_2 the rate of formation of CO_2 from n-butane and r_3 the rate of formation of CO_2 from maleic anhydride. The kinetic parameters are shown in Table 2.1.

Results of the computer simulations showed that the plug-flow model predicts higher conversions at higher gas velocities than the two-phase models while

Table 2.1 Empirical kinetic parameters (Moustafi et al. 2001)

Parameter	Value	Units
k_1	6.230×10^{-7}	$\text{mol}^{1-\alpha}\ \text{L}^{\alpha}/(\text{g s})$
k_2	9.040×10^{-7}	$\text{mol}^{1-\beta}\ \text{L}^{\beta}/(\text{g s})$
k_3	0.966×10^{-7}	$\text{mol}^{\gamma-\delta}\ \text{L}^{1-\delta-\gamma}\ (\text{g s})$
K_B	2616	mol/L
α	0.2298	
β	0.2298	
γ	0.6345	
δ	1.151	

conversion of n-butane and yield of maleic anhydride both decreased with increasing gas velocity. The results also showed the conversion to decrease with increasing n-butane feed concentration although the selectivity to maleic anhydride was higher at lower n-butane concentrations.

(ii) Dente et al. (2003) developed a bubbling-bed model to describe the operation of the reactor employed in the ALMA process referred to above. Figure 2.6 gives an outline of the reactor geometry and the structure of the reactor model. The model was based on one version of the two-phase model of Davidson and Harrison (1963) in which:

(a) emulsion-phase gas is completely mixed
(b) bubble-phase gas is in plug flow
(c) interphase mass transfer occurs by a combined process of molecular diffusion and throughflow
(d) reaction occurs only in the emulsion phase at an experimentally-determined rate.

Dente et al. first measured the reaction kinetics of the system by means of a tubular microflow reactor loaded with a commercial VPO catalyst, analysing the data on the basis of a scheme comprising five parallel reactions in which n-butane was oxidised to maleic anhydride, acrylic acid, acetic acid, CO and CO_2 and three consecutive reactions in which the anhydride and the two organic acids were oxidised to CO and CO_2.

The Davidson-Harrison model gives the reactant concentration in the emulsion phase as:

$$C_{Ae} = \frac{C_{A0}(1 - f_b e^{-K_{be}})}{1 - f_b e^{-K_{be}} - k'} \tag{2.4}$$

and in the bubble phase as:

$$C_{Ab} = C_{Ae} + (C_{A0} - C_{Ae})e^{-K_{be}} \tag{2.5}$$

where C_{A0} is the initial concentration of reactant, f_b the bubble-phase gas fraction, K_{be} the interphase mass-transfer coefficient (bubble to emulsion) and k' the appropriate reaction rate coefficient. Dente et al. calculated f_b from bubble sizes and gas velocities as reported by Kunii and Levenspiel (1991) and interphase mass-transfer coefficients (bubble-to-cloud, K_{bc}, cloud-to-emulsion, K_{be}) again from Kunii and Levenspiel (1991):

$$K_{bc} = 4.5\frac{u_{mf}}{d_b} + 5.85\left(\frac{D_{bc}^{1/2}g^{1/4}}{d_b^{5/4}}\right) \tag{2.6}$$

$$K_{ce} = 6.78\sqrt{\frac{D_{ce}^{1/2}u_b\varepsilon_{mf}}{d_b^3}} \tag{2.7}$$

where d_b is the bubble diameter, D_{bc} is the bubble-cloud diffusion coefficient, D_{ce} is the cloud-emulsion diffusion coefficient, g is gravity, u_b is the bubble rise velocity and ε_{mf} is the emulsion-phase voidage. The model was solved on the basis of the tanks-in-series structure shown in Fig. 2.6 although no indication is given of the number of tanks, N, considered (Fig. 2.6).
Model predictions were compared with daily plant data averaged over a two-year period and expressed in terms of n-butane conversion and maleic

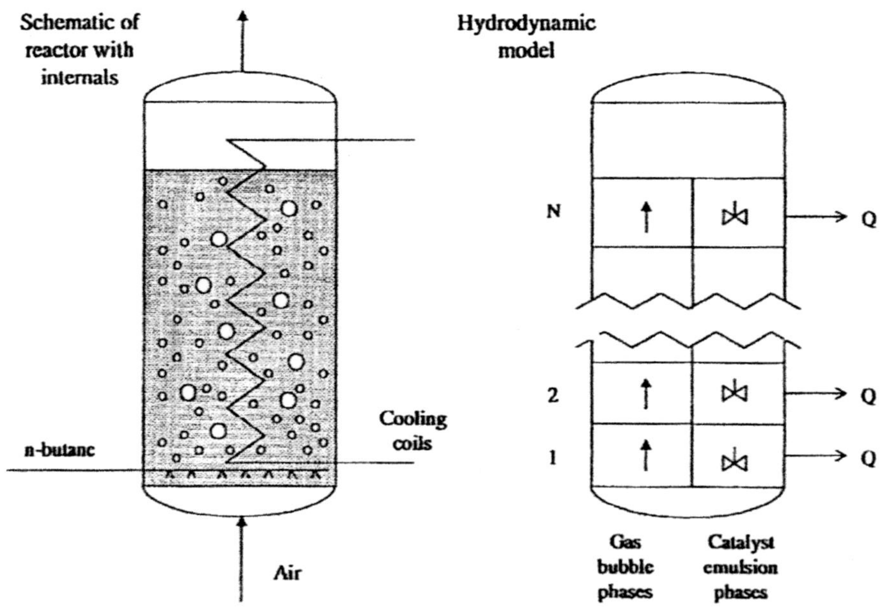

Fig. 2.6 Schematic of the fluidized-bed reactor (diameter 7 m) for the simulation of the ALMA process and discretization sections (Dente et al. 2003)

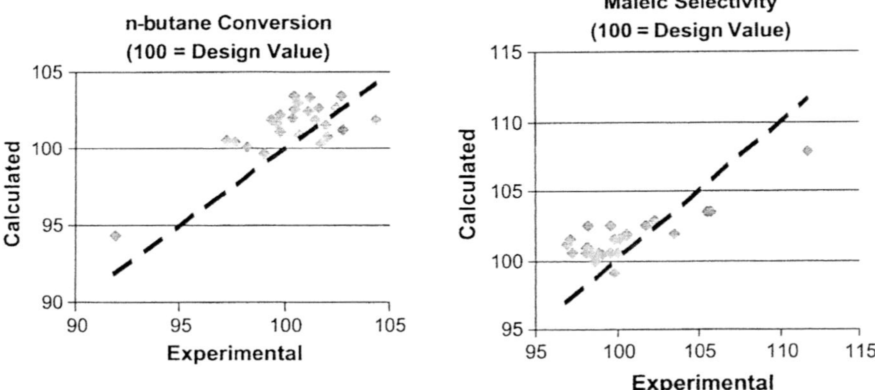

Fig. 2.7 Model predictions versus experimental values for n-butane conversion (Dente et al. 2003

anhydride selectivity, the results being shown in Fig. 2.7. It is clear that the model overestimates both values but the authors conclude that the discrepancies are acceptable.

(iii) Jiang et al. (2003) developed a turbulent-bed model and applied it to the catalytic oxidation of n-butane to maleic anhydride. The model was based on the two-phase concept with a dilute phase containing particles and a dense phase as before. The gas in both phases was assumed to be in plug flow. Since significant particle entrainment occurs in turbulent beds the authors applied a one-dimensional plug-flow model to describe reaction in the freeboard region. The underlying assumptions were (a) the turbulent-bed reactor operates isothermally; (b) catalyst activity is uniform and remains unchanged throughout; (c) reaction takes place in the dense and dilute phases as well as in the freeboard.

A number of authors have developed models of the riser/regenerator system such as that employed in the DuPont process.

(iv) Pugsley et al. (1992) modelled the riser hydrodynamics on the basis of a dense turbulent zone at the base where solids are introduced from the standpipe followed by a zone of fully–developed flow characterised by a core-annular structure with a lean core in which gas and solids flow vertically upwards and a denser gas-solid zone moving downwards at the wall. In view of the smooth exit of the riser to the external cyclone there was no deceleration zone to consider the fully-developed zone being assumed to extend all the way to the outlet. The details of the flow structure were derived from the earlier models of Berruti and Kalogerakis (1989) and Wong et al. (1992). The reaction kinetics incorporated into the model were those of Centi et al. (1985) referred to above. The assumptions made by Pugsley et al. in their computer simulations are listed in Table 2.2.

Table 2.2 Major assumptions for the computer simulations (Pugsley et al. 1992)

The CFB catalytic reactor operates isothermally
Reaction occurs in both the core and annular regions
Catalyst deactivation is the same in both core and annular regions at the same riser
Axial location
Gas input to the annular region is due solely to crossflow of gas from the core
Reaction is chemically controlled
Riser diameter = 0.3 m; riser height = 20 m
The riser is equipped with a smooth exit to the cyclone
G_s = 400–800 kg/(m^{2}s)
U_0 = 4–6 m/s
c_B = 1–50 mol%
T = 573 K
VPO catalyst physical properties (Geldart A)
D_p = 75 μm
ρ_s = 1500 kg/m^3
ε_{mf} = 0.5
U_t = 0.05 m/s

The riser was divided into 100 elements of 20 cm each and from the Berruti and Kalogerakis model core voidage and core radius were found for each element. Mass balance equations for the core and annular regions derived by Patience (1990) were solved for a given inlet concentration of reactants to the first volume element applying the kinetic data of Centi et al. Since the majority of gas was assumed to flow in the core region the reactant concentration in the annulus was assumed to be zero in the first element. The resulting outlet concentrations of reactants and products were then input to the next element and so on to the reactor outlet. The main conclusions of the work were:

- conversion to maleic anhydride decreased but selectivity increased with increasing n-butane feed concentration
- conversion increased at higher solid circulation rates due to increased solids hold-up and better gas-solid contacting
- conversion decreased with increasing gas superficial velocity
- catalyst deactivation had only a slight influence on reactor performance.

(v) Golbig and Werther (1997) carried out an experimental study of n-butane oxidation in a coupled riser-regenerator system using a specially-prepared microspheroidal VPO catalyst. The riser (21 mm i.d., 2880 mm length) was fed with a mixture of n-butane and nitrogen the entrained solids from the riser outlet passing to a stripper and then to the regenerator, a bubbling bed (51–102 mm i.d., 877 mm length) fluidized with an oxygen/air mixture. The freshly regenerated catalyst particles leaving the regenerator were passed to

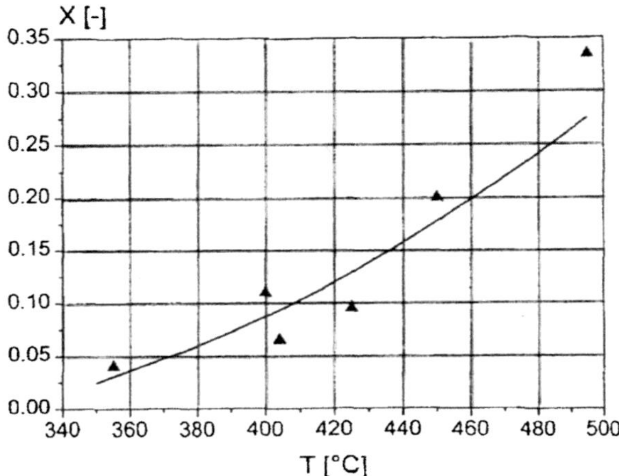

Fig. 2.8 Comparison of the n-butane conversion temperature dependence with model calculations (u_g = 2 m/s; 5 mol% n-butane in the riser; 12–15 mbar pressure drop in the riser; 50 mol% oxygen in the regenerator) (Golbig and Werther 1997)

the riser where they reacted with the n-butane feed to produce MAN, water, CO and CO_2. In the model formulation the riser was represented by two phases: the gas phase and the catalyst phase, mass transfer between the two being described by a previously developed correlation (Vollert and Werther 1994). The regenerator was modelled as a two-phase bubble-emulsion system. In both riser and regenerator the catalyst was assumed to be completely mixed while the gas in both units was in plug flow. Owing to limitations in the supply of catalyst the total operation time of the experimental facility was restricted to some 60 h but sufficient data were collected to enable comparisons to be made with the hydrodynamic model.

Figure 2.8 shows a comparison between n-butane conversion, X, over a range of riser temperatures and model predictions while Fig. 2.9 compares X for a range of n-butane concentrations with the model. The authors conclude

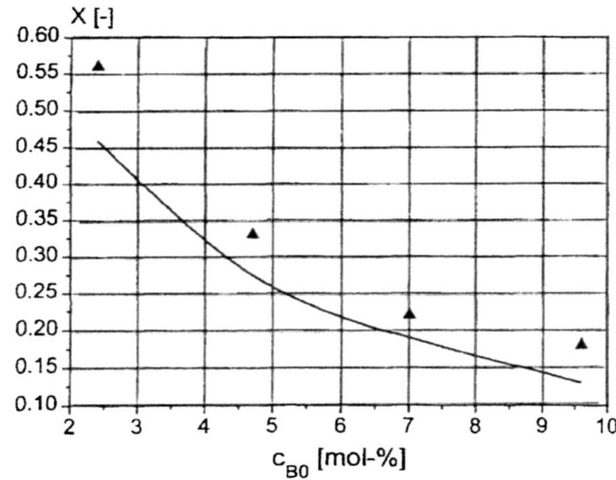

Fig. 2.9 Influence of n-butane feed concentration on the butane conversion—a comparison of measurements and model calculations (500 °C; u_g = 2 m/s; 12–19 mbar pressure drop in the riser; 50 mol% oxygen in the regenerator) (Golbig and Werther 1997)

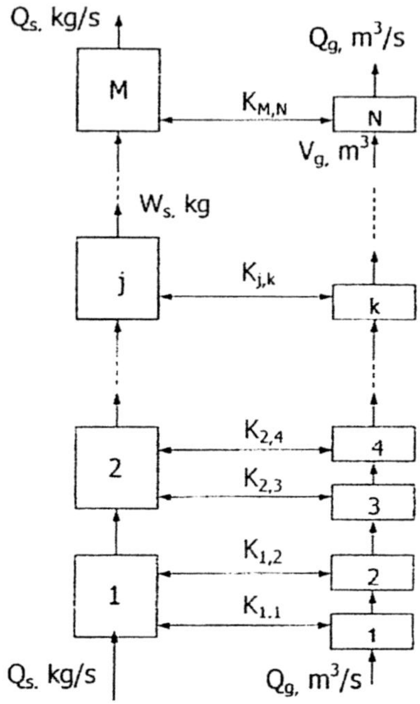

Fig. 2.10 Schematic diagram of the model formulation (Roy et al. 2000)

that the model gives an adequate description of trends in conversion and selectivity to MAN whilst acknowledging that improvements could be made by assuming a more realistic description of the riser in terms of a core-annular structure such as that assumed by Pugsley et al. (1992).

(vi) Roy et al. (2000) based their model on two kinetic schemes: (a) that of Centi et al. (1985) and (b) that of Mills et al. (1999) referred to above. In the model the reaction kinetics were decoupled from the riser hydrodynamics which were described in terms of a series of mixing cells for both solids (M cells) and gas (N cells) with interchange between both (Fig. 2.10). Simulation of a riser as described by Pugsley et al. (1992) was then carried out using the commercial CFD package FLUENT in which the solid and gas phases were assumed to be interpenetrating continua.

The model predictions are shown in Table 2.3 from which it is clear the reactor performance increases the closer the hydrodynamics approach plug flow and that both conversion and yield of maleic anhydride approach asymptotes as the number of compartments increases. The authors conclude that the model of Mills et al. (1999) is the more appropriate for this application since it was measured under conditions that mimic the cyclic exposure of the VPO catalyst to n-butane and oxygen.

In recent years the process simulation tool SolidSim has been developed by eleven institutes from nine German Universities specifically as a means of modelling processes involving fluids and solids (Hartge et al. 2006). The tool

Table 2.3 Effect of mixing pattern on n-butane conversion and MAN yield (Roy et al. 2000)

Number of solid compartments	Number of gas compartments[a]	Conversion of n-butane	Yield of MAN
1	1 (1)	73.3	19.3
	2 (2)	78.1	24.2
	3 (3)	78.6	24.7
	4 (4)	78.6	24.8
	5 (5)	78.6	24.8
	6 (6)	78.6	24.8
2	2 (1)	80.2	27.9
	4 (2)	82.7	28.1
	6 (3)	82.9	28.4
	8 (4)	83	28.5
3	3 (1)	85.9	29.8
	6 (2)	88	31.6
	9 (3)	88.6	32

[a]Numbers in parentheses show the number of gas phase compartments exchanging mass with each solid-phase compartment

was applied by Puettemann et al. (2012a, b) to the selective oxidation of n-butane to maleic anhydride in a riser/regenerator system on both laboratory and industrial scales. The laboratory simulations were based on the earlier work of Goldbig and Werther (1997) while the large-scale simulations used operating data from the DuPont plant referred to above. A major conclusion from the latter simulation was that the solids circulation rates needed to achieve the design yields were unrealistically high, a conclusion in keeping with experience on the plant where the catalyst was never sufficiently regenerated and to compensate it was necessary to inject additional oxygen into the bottom zone of the riser. In the original DuPont experimental unit 50 % of the oxygen came from the regenerator; in the pilot plant this was reduced to some 20 % while in the commercial unit it was often as little as 10 % (Personal Communication). As noted above this was one reason for the process being abandoned after some years of operation.

2.2.3 *Propylene Ammoxidation to Acrylonitrile*

2.2.3.1 Process Background

Acrylonitrile, a low-boiling (b.p. 77 °C) flammable material, is used extensively in the production of acrylic fibres and resins, ABS rubbers (acrylonitrile-butadiene-styrene) and speciality products. Worldwide production in

2002 was estimated to be 5 Mt/a (Brazdil 2005). It is produced almost exclusively by a fluidized-bed process invented in the late 1950's by the Sohio company (Idol 1959). The process was designated a National Historic Landmark by the American Chemical Society in 1996.

2.2.3.2 Sohio Process

This so-called ammoxidation process involves the reaction between propylene, ammonia and oxygen (air) over a solid catalyst:

$$CH_2{=}CHCH_3 + NH_3 + 3/2O_2 = CH_2{=}CHCN + 3H_2O; \Delta H = -515\,\mathrm{kJ/mol}$$

In addition, side reactions occur leading to the formation of HCN, acetonitrile, acrolein and oxides of carbon as a result of which the overall enthalpy of reaction is in the region of 670–730 kJ/mol (Kunii and Levenspiel 1991). HCN and acetonitrile are important co-products of the process and are separated from the product stream for further processing. Fluidized-bed operation with cooling coils immersed in the bed enables the heat of reaction to be controlled and bed temperatures to be maintained within the required range of 400–460 °C. Furthermore the flame-quenching action of the moving bed particles enables reactants and products to be processed without hazard despite the wide flammability limits of the organic components in air (Sax 1975).

A typical reactor layout is shown in Fig. 2.11.

Reactor diameters are in the range 3–8 m and are operated at 1.3–2 bar pressure. Air is fed into the unit through a bottom distributor while a mixture of propylene and ammonia enters through sparger pipes with downward-pointing orifices located below the in-bed cooling coils which are fed with water to generate high-pressure steam used to drive the air compressor and for downstream applications. The molar feed ratio of propylene/ammonia/air is 1:1.15:10 giving a minimum excess of ammonia over propylene with about 10 % excess air with respect to propylene (Jiang et al. 2003). To maintain a good quality of fluidization multiple internal cyclones maintain the bed-particle size in the range 10–200 μm with the proportion of fines (<44 μm) being 20–40 %. The oxygen-rich region between the air distributor and the sparger pipes serves to burn off carbon deposits on the catalyst and to reoxidize its surface thereby maintaining the lifetime of the catalyst for prolonged periods (Kunii and Levenspiel 1991). Gas velocities are in the range between 0.4 and 0.5 m/s indicating turbulent-regime flow. Trays or screens can be placed horizontally to reduce gas backmixing and so to improve performance. Gas residence time in the reactor is in the optimal range of 5–10 s resulting in almost 100 % per-pass conversions of propylene with selectivities to acrylonitrile of around 80 % (Dimian and Bildeac 2008).

The use of propane as an alternative feedstock to the more expensive propylene has been discussed in recent years in both the scientific literature (Centi et al. 1993; Fakeeha et al. 2000; Dimian and Bildeac 2008) and in patents (Glaeser et al. 1989).

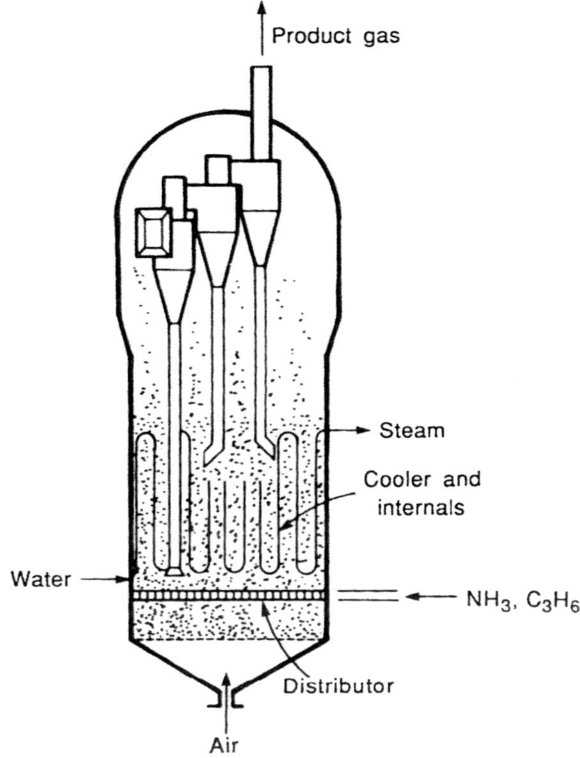

Fig. 2.11 Schematic of a Sohio ammoxidation reactor (Kunii and Levenspiel 1991)

Catalysts similar to those used in the Sohio process have been proposed, some containing vanadium and antimony, but selectivities to acrylonitrile have generally been lower at around 60 % with conversions of 50 %. This inferior performance coupled with the problem of higher temperature operation (500–550 °C) producing a wider range of by-products has proved insufficient to justify the massive replacement costs of existing technologies.

2.2.3.3 Catalysts

The catalyst employed in the original Sohio process was bismuth phosphomolybdate supported on microspheroidal silica particles in which the bismuth component initially activates the propylene molecule by abstraction of a hydrogen atom to give an adsorbed π-allyl intermediate. The function of the molybdenum is thought to be to activate the ammonia molecule to generate NH species which are then inserted into the π-allyl group to form acrylonitrile precursors. This is followed by rearrangement, additional hydrogen abstraction and desorption of the resulting acrylonitrile. The complex sequence of reactions has been described in detail by Grasselli (1986, 1999) and Hanna (2004). Catalysts currently in use are multicomponent materials containing a range of metals such as iron, nickel, cobalt, magnesium, caesium and potassium (Grasselli 1999). The iron acts as a redox couple ($Fe^{3+/2+}$) transferring lattice oxygen to the Bi–O–Mo active site. The Fe^{2+} surface sites are stabilised by divalent elements such as Ni, Co, Mg and Mn to form stable molybdates that are isostructural with Fe^{2+} molybdates. Alkali metals such as potassium serve to neutralise acidic cracking sites on the catalyst surface. These multicomponent formulations give superior performance in terms of activity and selectivity over the original version, in most cases giving an acrylonitrile yield in excess of 75 mol% based on propylene feed (Grasselli 1999).

2.2.3.4 Reactor Modelling

Kunii and Levenspiel (1991) considered the ammoxidation of propylene on the basis of the following reaction scheme:

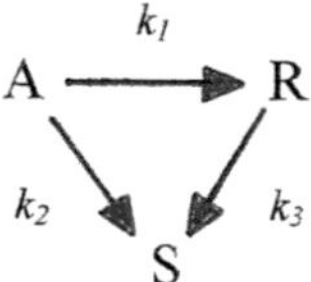

where A = propylene, R = acrylonitrile, S = HCN, CO, CO_2 etc., k_1, k_2 and k_3 being the corresponding reaction rate coefficients. The system was modelled using the authors' "bubbling bed" model (Kunii and Levenspiel 1969) and calculations were carried out to determine the dimensions and operating parameters of a commercial-scale reactor packed with an array of vertical cooling tubes. The model gives the fraction of propylene unconverted leaving the reactor as:

$$\frac{C_A}{C_{A0}} = \exp\left(-k_{f12}\tau\right) \tag{2.8}$$

and the conversion, X_A as:

$$X_A = 1 - \frac{C_A}{C_{A0}} \tag{2.9}$$

where k_{f12}, the effective rate coefficient for propylene conversion, is given by:

$$k_{f12} = \frac{\delta}{1-\varepsilon_f}\left[\gamma_b k_{12} + \frac{1}{\frac{1}{K_{bc}} + \frac{1}{\gamma_c k_{12} + \frac{1}{\frac{1}{K_{ce}} + \frac{1}{\gamma_e k_{12}}}}}\right] \tag{2.10}$$

where: $k_{12} = k_1 + k_2$

K_{bc} = bubble-to-cloud mass transfer coefficient
K_{ce} = cloud-to-emulsion mass transfer coefficient
γ_b = fraction of solids in bubbles
γ_c = fraction of solids in the cloud
γ_e = fraction of solids in the emulsion
δ = bubble volume fraction
ε_f = emulsion-phase voidage

K_{bc} and K_{ce} are found from Eqs. (2.6) and (2.7). u_0 and u_b are the superficial gas velocity and the bubble rise velocity respectively while:

$$\delta = \frac{u_0}{u_b} \tag{2.11}$$

$$\delta(\gamma_b + \gamma_c + \gamma_e) = (1 - \varepsilon_{mf})(1 - \delta) \tag{2.12}$$

and

$$\gamma_b, \gamma_c, \gamma_e = \frac{volume\ of\ solids\ dispersed\ in\ b, c, e\ respectively}{volume\ of\ bubble} \tag{2.13}$$

In Eq. (2.8) τ is a gas residence time defined as:

$$\tau = \frac{L_f(1 - \varepsilon_f)}{u_0} \tag{2.14}$$

where L_f is the bed height, and the average bed voidage, ε_f is:

$$\varepsilon_f = \delta + (1 - \delta)\varepsilon_e \tag{2.15}$$

ε_e being the emulsion-phase voidage.

The model was solved for a number of values of the reaction rate coefficients, conversions and selectivities being plotted in Fig. 2.12 the results then being applied to calculate the overall dimensions of a commercial-scale unit to produce 50,000 tons of acrylonitrile per 334-day year with a propylene conversion of 95 % and a selectivity to acrylonitrile of at least 65 %. The heat of reaction was to be removed via an array of vertical heat exchanger tubes 0.08 m in diameter and 7 m in length that controlled the bed temperature at 460 °C; the bed pressure was to operate at a pressure of 2.5 bar. Applying heat-transfer correlations from the literature the required number of tubes was calculated to be 296 arranged in a square array with a pitch of 0.323 m the resulting bed diameter being 7.18 m.

The calculations are a good example of the application of a fluidized-bed reactor model to the design of a full-scale reactor and of the physical and chemical data required to carry out the design.

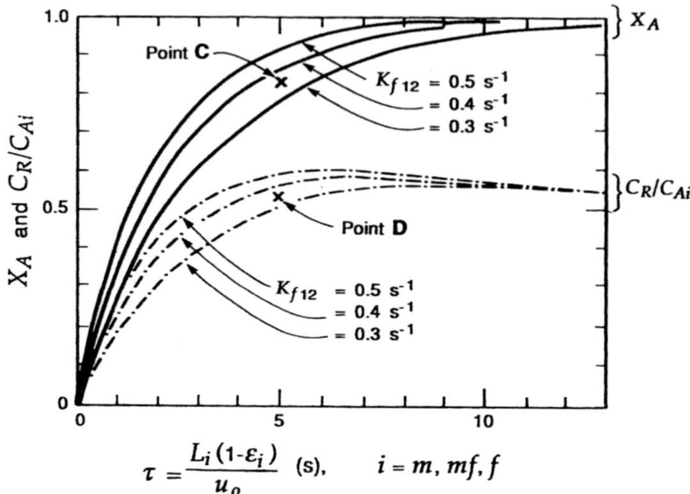

Fig. 2.12 Comparison of calculated lines with experimental data for propylene ammoxidation (Kunii and Levenspiel 1991)

2.2.4 Vinyl Chloride Monomer (VCM)

Vinyl chloride, the precursor material for polyvinyl chloride (PVC), is made by the thermal cracking of ethylene dichloride (1,2-dichloroethane) in a tubular reactor at 450–600 °C and 10–35 bar:

$$CH_2ClCH_2Cl \rightarrow CH_2{=}CHCl + HCl; \Delta H = 72.6\,kJ/mol$$

The ethylene dichloride itself is made by (a) the direct chlorination of ethylene in the liquid phase with itself as solvent and ferric chloride (Fe(III) chloride) as catalyst at 40–100 °C and 1–10 bar:

$$CH_2{=}CH_2 + Cl_2 \rightarrow CH_2ClCH_2Cl;\ \Delta H = -180\,kJ/mol$$

and (b) the oxychlorination reaction in which HCl recovered from the pyrolysis stage is reacted with ethylene, air or oxygen in a fluidized-bed reactor at 220–245 ° C and 2.5–6 bar:

$$CH_2{=}CH_2 + HCl + {}^1\!/_2O_2 = ClCH_2CH_2Cl + H_2O; \Delta H = -237.3\,kJ/mol$$

Combining the three processes of direct chlorination, ethylene dichloride pyrolysis and oxychlorination leads to the so-called “balanced” process with no net consumption or production of HCl. The highly exothermic oxychlorination reaction is catalysed by alumina-supported copper oxide and strict temperature control is necessary to prevent (i) catalyst agglomeration, (ii) ethylene combustion, (iii) over chlorination of the ethylene dichloride product. The catalyst is a Geldart Group A material, the fluidized-bed reactor having many features in common with the Sohio ammoxidation reactor shown in Fig. 2.11 (Jazayeri 2003). Isothermal operation is achieved by the use of densely-packed serpentine cooling coils immersed in the bed the material of construction of both reactor and coils being carbon steel (Bolthrunis et al. 2004).

2.2.5 Vinyl Acetate Monomer (VAM)

Vinyl acetate, a major industrial chemical with worldwide production in excess of 4.5 Mt/a, is used in the production of polyvinyl acetate, polyvinyl alcohol and a number of other materials for use as adhesives, films and emulsion-based paints. The main production route for VAM is the acetoxylation reaction between ethylene, acetic acid and molecular oxygen in fixed-bed reactors over a catalyst containing palladium, gold and a promoter such as potassium acetate supported on silica:

$$CH_2{=}CH_2 + CH_3CO_2H + {}^1\!/_2\,O_2 \rightarrow CH_2{=}CHCO_2CH_3 + H_2O;\ \Delta H = -211.2\,\mathrm{kJ/mol}$$

A typical catalyst of the type introduced by Bayer in the 1960’s consists of 0.5–1.5 % Pd, 0.2–1.5 % Au, 4–10 % KOAc on silica in the form of 5 mm diameter spheres operated at 140–180 °C and 5–12 bar pressure.

In late 2001 BP Chemicals, in an attempt to establish a new technology in the field, started up the world’s first fluidized-bed process for VAM at their Saltend, Hull facility in the UK with a capacity of 250 kt/a from a single reactor. The advantages of the fluidized-bed for the process were simplicity of design, increased catalyst life (since catalyst deactivation was minimised by the absence of hot spots common in fixed beds), continuous addition of make-up catalyst, and higher production rates since higher oxygen levels could be used without the risk of forming a flammable mixture in the fluidized environment. The Pd/Au-based catalyst was a microspheroidal Geldart Group A material (Baker et al. 2003) in a reactor similar in form to that referred to above for the oxychlorination process for VCM (Fig. 2.17). The plant was operated successfully for a number of years with no reported technical problems. However having been acquired from BP by INEOS in 2008 and owing to a combination of the availability of low-cost imports and a hostile trading environment the process became uneconomic and the plant was shut down in October 2013.

2.2.6 *Gas-to-Liquid Technologies*

Factors such as the volatility of the international oil market and the lack of indigenous sources of crude oil in various countries have prompted the development of processes to produce liquid fuels and chemicals from coal and natural gas. Thus in Germany in the period between the two World Wars two processes, the Bergius process for the production of hydrocarbon fuels by the high-pressure hydrogenation of brown coal and the Fischer-Tropsch process for generating liquid fuels and chemicals from synthesis gas, were developed and commercialised. In the Bergius process finely divided coal was slurried with recycled oil containing an iron catalyst and hydrogenated at 400–500 °C and 20–70 MPa pressure to give a synthetic crude oil. The Bergius process is no longer practiced but variants of the Fischer-Tropsch process have proliferated in countries such as South Africa and China with widespread deposits of coal but little in the way of petroleum or natural gas.

Synthesis gas or "syngas" is made by the steam gasification of coal using, for example, Lurgi dry-ash gasifiers (Dry 1996). Lurgi gasifiers operate under pressure (2–3 MPa) and use a steam/oxygen mixture as the gasifying medium:

$$3C + O_2 + H_2O \rightarrow H_2 + 3CO$$

The feedstock is lump coal which is admitted to the reactor via a pressurised hopper and kept in motion during reaction by means of a rotating grate through which the ash is discharged. The crude syngas is a mixture of mainly H_2, CO, CO_2 and CH_4 the actual compositions depending on the process conditions and the type of coal used. For use in the Fischer-Tropsch process the purified syngas composition should be such that the ratio $H_2/(2CO + 3CO_2)$ is slightly greater than 1.0 (Dry 1996). Where necessary the H_2/CO ratio may be adjusted via the nickel-catalysed water-gas shift reaction:

$$H_2O + CO \leftrightarrow H_2 + CO_2;\ \Delta H = -41\ \mathrm{kJ/mol}$$

The Fischer-Tropsch reaction may be represented by:

$$(2n + 1)H_2 + nCO \rightarrow C_nH_{2n+1} + nH_2O$$

where n is in the range 10–20. The alkane products are largely linear and vary from methane to heavy waxes; the liquid products are suitable as diesel fuel but the gasoline yield is low and of poor quality.

2.2.6.1 Synthol Process

The South African Sasol company have since the mid-1950's operated large-scale units for the production of a range of coal-derived gases and liquids via a combination of the Lurgi and Fischer-Tropsch processes. In the early years a total of thirteen Lurgi gasifiers were installed each of some 4 m diameter with a total capacity of around 8.2×10^6 m^3/day of raw gas. The gasifiers supplied gas to two Fischer-Tropsch variants, the Arge process and the Synthol process. The Arge plant consisted of five packed-bed reactors producing 18,000 tonnes/year of hydrocarbons. The Synthol plant had two circulating fluidized-bed (CFB) reactors each with a capacity of 65,000 tonnes/year (2200 bbl/d) of hydrocarbons. The original Synthol reactors were designed on the basis of data from 4 and 10 cm diameter pilot-plant units obtained by the Kellogg company in the USA and the decision was taken to build CFB reactors rather than the alternative bubbling beds that had been used unsuccessfully in the previously-mentioned Hydrocol process (Sect. 1.2.1.2). The decision to build CFB reactors "although causing much anguish initially, paid off handsomely in the end" (Duvenhage and Shingles 2002). Here the powdered (Geldart A) iron catalyst is carried upwards in dilute-phase flow by the fluidizing syngas at 3–12 m/s and temperatures initially of around 315 °C but rising as the exothermic synthesis reactions start to occur. Heat is removed in coolers situated in the expanded section of the riser the temperature reaching a maximum of 350 °C in the hopper above the standpipe. After separation of product gases and catalyst in banks of cyclones the catalyst is returned via the hopper and standpipe to be picked up by the incoming syngas to renew the process.

As a consequence of the embargo on the import of oil-based products into South Africa imposed by the OPEC in the early 1970's the Sasol company in association with Badger embarked on a major expansion of the coal-to-oil technology. The result was the construction of eight enormous 60 m tall CFB reactors that started up in 1980; a further eight such units began operation in 1982; the overall dimensions of the reactors are shown in Fig. 2.13.

The CFB reactors and their complex support structures were costly to build and expensive to operate and maintain and during their period of development the company began work on the design of a less expensive system based on the so-called Sasol Advanced Synthol (SAS) turbulent dense-phase fluidized-bed reactor. The first commercial-scale SAS unit based on a 5 m diameter 3000 bbl/d reactor began operation in 1989 followed by an 8 m diameter 11,000 bbl/d unit two years later (Sookai et al. 2001). During 1998/99 the 16 CFB reactors were replaced by four 10.7 m and four 8 m diameter SAS reactors (Duvenhage and Shingles 2002); the dimensions and product yields of the reactors are shown in Fig. 2.13.

Synthol catalyst is a promoted iron powder (Geldart Group A) derived from the millscale produced in steel making. In addition to the Fischer-Tropsch reaction shown above the catalyst promotes a side reaction, the Boudouard reaction, in which CO is decomposed to CO_2 and carbon the latter reacting to form iron carbide on the catalyst surface. This results in a decrease in the particle density of the catalyst which if unchecked would lead to uncontrollable bed expansion and catalyst loss through the cyclones. To counter this carbided catalyst is removed continuously and replaced with fresh material giving an optimum balance between the good flow properties of the used material and the high conversion potential of the fresh (Sookai et al. 2001).

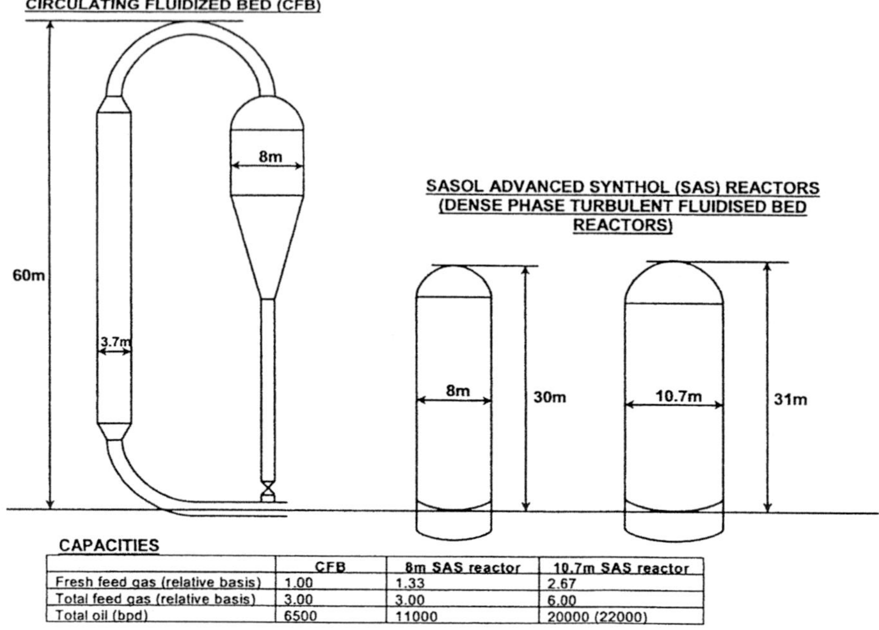

	CFB	8m SAS reactor	10.7m SAS reactor
Fresh feed gas (relative basis)	1.00	1.33	2.67
Total feed gas (relative basis)	3.00	3.00	6.00
Total oil (bpd)	6500	11000	20000 (22000)

Fig. 2.13 Size and capacity comparison between Synthol CFB and SAS reactors (Sookai et al. 2001)

Some of the advantages of the SAS units over the CFB units have been set out by Duvenhage and Shingles (2002) as follows:

- higher per-pass conversions
- lower catalyst consumption (by 50 % per ton) of product
- excellent isothermal performance and temperature control
- less maintenance
- greater run stability
- less erosion of critical components and less catalyst attrition

2.2.6.2 Methanol-Based Processes

Driven by the volatility of the market in petroleum products in the 1970’s the Mobil company (now ExxonMobil) developed a process for converting methanol to gasoline; the MTG process was first commercialised at Motunui in New Zealand in the mid-1980’s. The route to methanol started with the steam reforming of methane in fixed bed reactors over a nickel-based catalyst followed by the water gas shift reaction:

$$CH_4 + H_2O \leftrightarrow CO + 3H_2; \quad \Delta H = +206\,kJ/mol$$
$$CO + H_2O \leftrightarrow CO_2 + H_2; \quad \Delta H = -41\,kJ/mol$$

The product gases were then converted to methanol in fixed beds over a copper catalyst:

$$CO_2 + 2H_2 \leftrightarrow CH_3OH; \quad \Delta H = -92\,kJ/mol$$

the resulting methanol then being dehydrated in a fixed-bed reactor over an alumina catalyst to give an equilibrium mixture of dimethyl ether, methanol and water:

$$2CH_3OH \leftrightarrow CH_3OCH_3 + H_2O$$

The mixture was passed to a series of fixed-bed reactors containing the zeolite catalyst ZSM-5 where it was mixed with recycle gas and converted to a mixture of hydrocarbons and water. The ZSM-5 structure has pores of diameter 5.1–5.6 Å leading to a hydrocarbon product in the C_1–C_{10} range which after downstream treatment yielded gasoline with an octane rating of 92–95. In the MTG stage the catalyst became deactivated by the deposition of coke and was regenerated by burning off with air. The original plant used five swing reactors with one being regenerated off-line at any one time the other four being run in parallel. For economic reasons the Moturui plant was shut down in 1996 but second generation versions of the process (Harandi 1993) have subsequently been introduced and are of particular interest in developing countries such as China which, like South Africa, have large deposits of coal but little crude oil.

A further development of the MTG process is the methanol-to-olefin (MTO) technology being introduced as a source of ethylene and propylene for the burgeoning polymer market. The MTO process originated with the discovery by Union Carbide of a new class of zeolite catalyst, the silicoaluminophosphates (SAPO, particularly SAPO-34) which showed high selectivity to light olefins from methanol. SAPO-34 is made up of narrow pores of diameter 5.1–5.6 Å connected to large cages and leading to a narrow product distribution in the C_1–C_5 range. A number of versions of the process are currently available commercially and some such as the UOP/Hydro MTO process employ fluidized-bed reactor/generator technology for catalyst management (Funk et al. 2013; Chen et al. 2005). Developments of MTG and MTO installations in China have recently been summarised by Minchener (2014). Academic interest in MTO catalysis has been considerable with groups worldwide reporting mechanistic, kinetic and modelling studies (Park and Froment 2004; Gayubo et al.. 2005; Zhou et al. 2008; Kaarsholm et al. 2010).

2.2.7 Fluidized Catalytic Cracking (FCC)

2.2.7.1 Process Background

Since its introduction over 70 years ago fluidized catalytic cracking has become the most widespread fluidized-bed process and arguably the most important catalytic process in all industry. Its aim is to convert low-value heavy petroleum distillate fractions into lighter high-value products boiling in the gasoline and light-oil ranges. In terms of throughput catalytic crackers are second only to the atmospheric distillation units on most refineries. On refineries the heavy residue from the primary atmospheric distillation unit is fed to the vacuum distillation column giving vacuum gas oil and a residue boiling in the range 340–560 °C and it is this residue that was the traditional FCC feedstock; modern units however can process a wide variety of feeds including hydro-treated gas oils and deasphalted oils, as well as atmospheric residue (Chen 2003). The primary function of FCC units is to produce gasoline, some 45 % of worldwide production coming either directly or from downstream units such as alkylation plant; middle distillate, light petroleum gas and light olefins for petrochemical conversion are also important products.

Over the years many engineering companies such as UOP, Stone and Webster and Kellogg Brown Root as well as oil companies such as Shell and Exxon/Mobil have developed different versions of FCC systems (Gary and Handwerk 2001) an example being shown in Fig. 2.14.

Although differing in configuration all these versions are based on the same general principles. Cracking is carried out in a vertical riser-reactor at 480–540 °C where hot catalyst is fluidized and carried upwards in plug flow by preheated vaporized feed introduced through atomising nozzles, the catalyst-to-oil ratio normally being in the range 4:1 to 9:1 by weight (Sadeghbeigi 2012). The endothermic cracking reactions occur in the riser over a two-to-four second period and in the process the catalyst becomes deactivated by the deposition of coke (typically 0.4–2.5 wt%) on its surface. Deactivated catalyst flows from the riser into the reactor which acts as a disengaging space and a housing for single- or two-stage cyclones. These deliver the catalyst to a stripper section at the base of the reactor where it is fluidized with steam in a bubbling-bed mode, the stripped vapours passing out overhead with the primary products. Separated catalyst then flows via a standpipe into the regenerator where it is fluidized with air, the coke is burned off at temperatures of 675–730 °C to a typical level of 0.05 wt% and its catalytic activity restored. The cycle of events is completed as regenerated catalyst flows from the regenerator into the riser to resume the process. The flow rate of the catalyst between regenerator and riser is normally regulated by means of a slide- or plug-valve which controls the pressure head necessary for catalyst circulation around the system. Coke combustion raises the catalyst temperature to that required in the cracking reactions so maintaining a heat balance around the system. Cracked products and steam leave from the top of the reactor and pass to a fractionator for separation into four product groups: light gases (C_1–C_4), gasoline (C_5–220 °C),

Fig. 2.14 A modern riser cracker

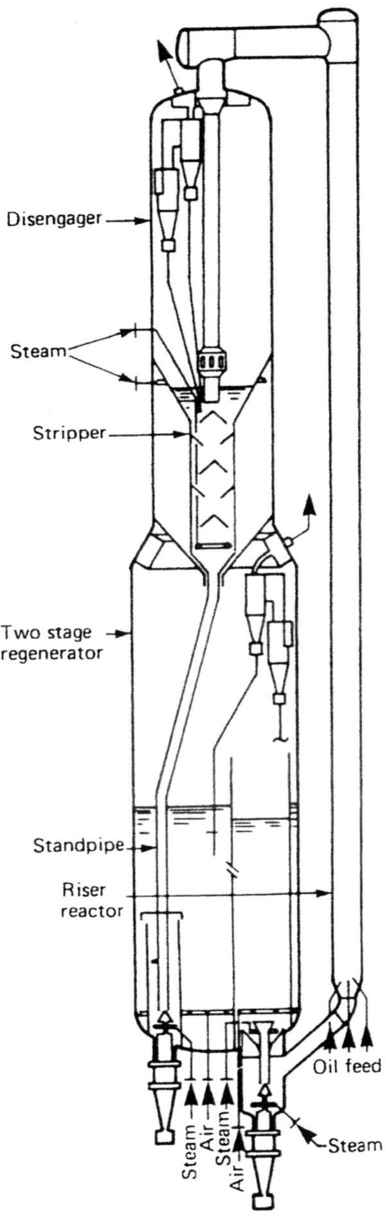

light cycle oil and heavy cycle oil (220–340 °C) and so-called slurry- or decant-oil (340 °C+). Steam and oxides of carbon, sulphur and nitrogen leave from the regenerator. By controlling the flow rates of feed, air and steam a continuous circulation of catalyst is maintained between riser-reactor and regenerator. In modern designs the risers operate in the fast fluidization regime at their lower end and, owing to the gas expansion caused by the cracking reactions, in dilute transport flow at their upper end; regenerators operate in turbulent-to-fast flow.

2.2.7.2 FCC Catalysts

Modern materials consist of four components: a zeolite, an amorphous silica-alumina matrix, a binder and a filler. The zeolite is the most active component and constitutes 15–50 wt% of the mixture. Zeolites are porous, crystalline aluminosilicates many of which have the general formula $M_{x/n}[(AlO_2)_x(SiO_2)_y] \cdot zH_2O$ where x/n is the number of exchangeable cations, M, of valency n. Those used in catalytic cracking have the structure of the naturally-occurring mineral faujasite but are chemically distinct from it. The basic unit of the faujasite structure is the cubo-octahedron made up of twenty four tetrahedra of either SiO_4^{4-} or AlO_4^{5-} connected through their hexagonal faces to give a unit cell of 192 tetrahedra with an edge length of 24.95 Å. It has an open, cage-like structure with large cavities interconnected by channels of 8–10 Å diameter allowing only smaller molecules to enter. One synthetic form, zeolite X, is made by crystallizing a sodium alumino silicate gel prepared by mixing aqueous solutions of sodium aluminate, sodium silicate and sodium hydroxide. The amorphous gel so formed is crystallized by agitation and heating at 100 °C and the sodium component ion-exchanged with other cations (NH_4^+, Ca^{2+}, La^{3+} etc.) to give the final structure. The Bronstedt- and Lewis-acid sites formed within the cavities are the active centres for the cracking reactions. The amorphous silica-alumina component of the catalyst promotes the cracking of larger hydrocarbon molecules while the binder and filler provide physical integrity and mechanical strength (Chen 2003).

Zeolites containing trivalent rare-earth cations such as La^{3+}, Ce^{3+}, Pr^{3+}, Nd^{3+}, Sm^{3+} are stable at high temperatures and in the presence of steam, rare earth exchanged materials containing less than 0.5 wt% Na_2O being able to withstand high steam concentrations up to 760 °C and thermal treatment up 815 °C. Of even greater importance than their increased activity compared with silica alumina is their greater selectivity to gasoline and their correspondingly lower yields of C_1–C_4 gases and coke (Table 2.4). This is attributed to their higher activity for hydrogen-transfer reactions relative to cracking in their small pore structure.

Of vital importance in determining the flow characteristics of FCC catalysts are particle size distribution (PSD) and the content of fines in the size range <40 μm. PSD's are in the range 10–150 μm with an average of around 70 μm i.e. a typical Geldart Group A material. Maintaining a steady concentration of fines is essential in preserving the flowability of the catalyst, any significant loss through cyclone malfunction seriously inhibiting catalyst circulation rates.

A number of materials act as poisons for cracking catalysts. Nitrogen compounds react with the acid centres and lower catalytic activity while the metal components of heavy oil fractions such as iron, nickel and vanadium deposit on

Table 2.4 Comparison of yield structure for fluid catalytic cracking of waxy gasoil over commercial equilibrium zeolite and amorphous catalysts (Venuto and Habib 1979)

Yields at 80 vol.% conversion	Amorphous high alumina	Zeolite XZ-25	Δ change from amorphous
Hydrogen wt%	0.08	0.04	−0.04
$C_1 + C_2$s wt%	3.8	2.1	−1.7
Propylene vol.%	16.1	11.8	−4.3
Propane vol.%	1.5	1.3	−0.2
Butenes vol.%	12.2	7.8	−4.4
i-Butane vol.%	7.9	7.2	−0.7
n-Butane vol.%	0.7	0.4	−0.3
Gasoline vol.%	55.5	62.0	+6.5
Light fuel oil vol.%	4.2	6.1	+1.9
Heavy fuel oil vol.%	15.8	13.9	−1.9
Coke wt%	5.6	4.1	−1.5
Gasoline octane number	94	89.8	−4.2

catalyst surfaces and increase the formation of gas and coke and reduce the yield of gasoline.

2.2.7.3 Process Chemistry

Hydrocarbon cracking over zeolite catalysts proceeds in the main by endothermic reactions involving carbenium-ion intermediates. Paraffinic molecules crack to produce olefins and smaller paraffins, and cycloparaffins (naphthenes). Aromatic compounds with alkyl side chains are either dealkylated completely to an unsubstituted aromatic and an olefin or partially cracked to a paraffin and an alkenyl aromatic. These primary reactions are followed by secondary processes leading to the final products. Important among these are hydrogen-transfer reactions, say from a naphthene to an olefin giving an aromatic and a paraffin, isomerizations forming iso-paraffins, and condensation reactions of aromatic residues leading to complex polynuclear hydrocarbons and ultimately to coke. Primary carbenium ions tend to isomerise to the more stable secondary and tertiary ions giving the cracked products a high concentration of highly branched molecules.

The cracking of a linear paraffin may be represented as follows.

Reaction is initiated by an interaction between an adsorbed hydrocarbon molecule and a proton from a Bronstedt acid site on the catalyst surface (Fig. 2.15i). Chain propagation follows by the ethyl carbenium ion abstracting a hydride ion from a second paraffin molecule (Fig. 2.15ii). The secondary carbenium ion so formed undergoes a β-scission reaction to form a primary carbenium ion and an olefin (Fig. 2.15iii). The primary carbenium ion formed in Fig. 2.15(iii) may propagate the chain by abstracting a hydride ion from another paraffin molecule as in Fig. 2.15(ii) or it may rearrange to form a more stable secondary ion

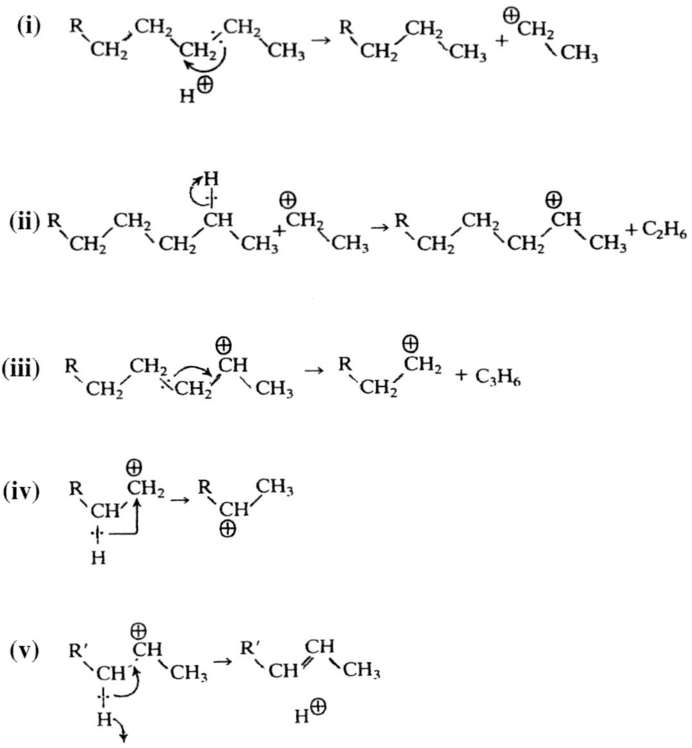

(i) $R{-}CH_2{-}CH_2{-}CH_2{-}CH_2{-}CH_3 \xrightarrow{H^{\oplus}} R{-}CH_2{-}CH_2{-}CH_3 + {}^{\oplus}CH_2{-}CH_3$

(ii) $R{-}CH_2{-}CH_2{-}CH_2{-}CH(H){-}CH_3 + {}^{\oplus}CH_2{-}CH_3 \rightarrow R{-}CH_2{-}CH_2{-}CH_2{-}\overset{\oplus}{C}H{-}CH_3 + C_2H_6$

(iii) $R{-}CH_2{-}CH_2{-}CH_2{-}\overset{\oplus}{C}H{-}CH_3 \rightarrow R{-}CH_2{-}\overset{\oplus}{C}H_2 + C_3H_6$

(iv) $R{-}CH(H){-}\overset{\oplus}{C}H_2 \rightarrow R{-}\overset{\oplus}{C}H{-}CH_3$

(v) $R'{-}CH(H){-}\overset{\oplus}{C}H{-}CH_3 \rightarrow R'{-}CH{=}CH{-}CH_3 + H^{\oplus}$

Fig. 2.15 Mechanism of catalytic cracking (Yates 1983)

(Fig. 2.15iv) which may then react as before. Chain termination occurs through donation of a proton back to the catalyst surface (Fig. 2.15v). Reactions of cycloparaffins follow a similar pattern.

The chemistry of catalytic cracking reactions was reviewed extensively by Venuto and Habib (1979) and by Sadeghbeigi (2012).

2.2.7.4 Operating Conditions and Heat Balance

The independent variables that define the reactor-regenerator operating conditions are reactor temperature, feed temperature, space velocity, catalyst activity and reactor pressure. The feed rate and air flow rate to the regenerator are set by flow controllers. The feed temperature is set by the fed temperature controller. Reactor temperature is controlled by the regenerator slide valve regulating the catalyst circulation rate (Chen 2003). The most important dependent variables are regenerator temperature, catalyst-to-oil ratio (i.e. catalyst circulation rate) and overall conversion. In the normal adiabatic mode of operation the combustion of coke on the catalyst provides the total heat requirement of the system. A simplified heat

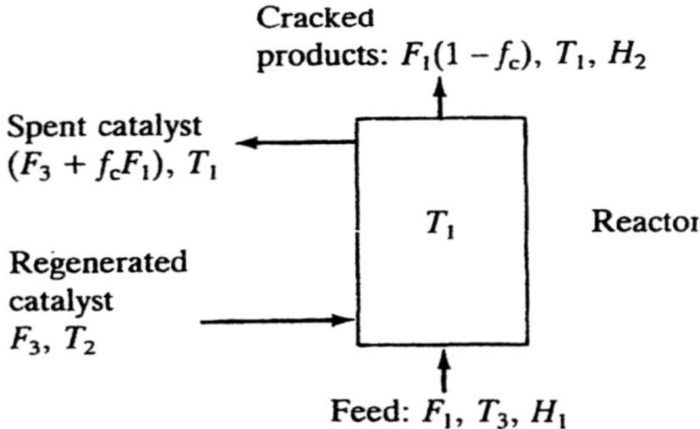

Fig. 2.16 Reactor heat balance (Yates 1983 adapted from Kunii and Levenspiel 1969)

where F_1 = **flow rate of feed (kg/s)**
F_3 = **flowrate of solid catalyst (kg/s)**
f_c = **mass fraction of feed deposited as coke**
T_1 = **reactor temperature (°C)**
T_2 = **regenerator temperature (°C)**
T_3 = **feed temperature (°C)**
H_1 = **feed enthalpy (kJ/kg)**
H_2 = **product enthalpy (kJ/kg)**

balance on the reactor (ignoring the effects of steam injection in the stripper and losses due to conduction and radiation) gives the following based on Kunii and Levenspiel (1969).

From Fig. 2.16:

$$\begin{aligned}&(\text{heat lost by catalyst}) = (\text{heat of cracking reaction}) + (\text{heat gained by feed})\\&F_3 C_{p,s}(T_2 - T_1) = F_1 \Delta H_{crac} + [F_1(1 - f_c)]H_2 - F_1 H_1\end{aligned} \tag{2.16}$$

$$\text{i.e. } \frac{F_3}{F_1} = \frac{\Delta H_{crac} + (1 - f_c)\,H_2 - H_1}{C_{p,s}(T_2 - T_1)} \tag{2.17}$$

where $C_{p,s}$ = specific heat capacity of solid catalyst (kJ/kg °C)

ΔH_{crac} = heat of endothermic cracking reaction (kJ/kg feed)

From Fig. 2.17:

$$\begin{aligned}&(\text{heat released by coke combustion}) = (\text{heat gained by gases}) + (\text{heat gained by catalyst})\\&-\Delta H_{comb} f_c F_1 = [(F_2 + f_c F_1)H_4 - F_2 H_3] + F_3 C_{p,s}(T_2 - T_1)\end{aligned} \tag{2.18}$$

ΔH_{comb} is the heat of the exothermic coke combustion reaction (kJ/kg coke).

A heat balance on the unit as a whole gives:

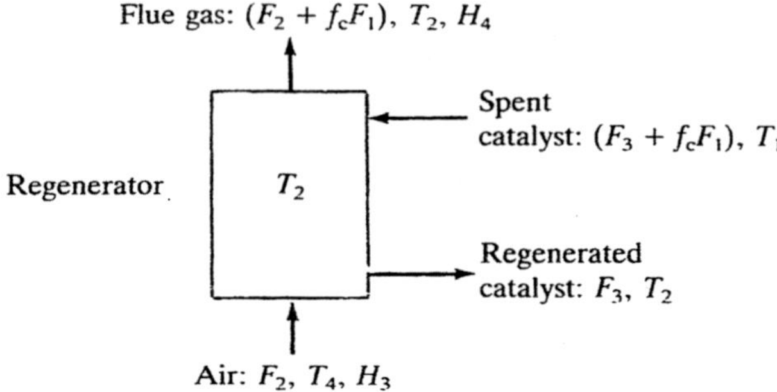

where F_2 = flowrate of air (kg/s)
T_4 = air temperature (°C)
H_3 = air enthalpy (kJ/kg)
H_4 = flue gas enthalpy (kJ/kg)

Fig. 2.17 Regenerator heat balance (Yates 1983 adapted from Kunii and Levenspiel 1969)

(heat of coke combustion) = (heat of cracking reaction) + (heat gained by feed)
+ (heat gained by gases)

from which it follows that:

$$\frac{F_2}{F_1} = \frac{-\Delta H_{comb} - \Delta H_{crac} + [H_1 - (1 - f_c)H_2] - f_cH_4}{H_4 - H_3} \tag{2.19}$$

Assuming the stoichiometry of the combustion reaction to be:

$$C + O_2 = CO_2 \quad \Delta H = -404.0\,\text{kJ/mol}$$

$$\frac{f_cF_1}{12} = \frac{0.21F_2}{1.293 \times 22.4}$$

$$\text{i.e. } \frac{F_2}{F_1} = 11.493 f_c \tag{2.20}$$

Since two of the primary variables in these equations, regenerator temperature and catalyst-to-oil ratio, F_3/F_1, are dependent variables the heat balance calculation requires a trial-and-error approach. For an assumed regenerator temperature, T_2, the catalyst coke content, f_c, may be found from Eqs. (2.19) and (2.20) for given values of H_1, H_2, H_3, T_1, T_3 and T_4. The catalyst-to-oil ratio may then be found from Eq. (2.17) and substituted into Eq. (2.18) and the calculated regenerator temperature compared with the value assumed initially. The highly coupled nature of the reactor-regenerator system is apparent from the foregoing analysis.

Catalyst circulation is determined by the pressure balance around the unit, the rate of circulation being regulated by two slide valves one on the stripper and one on the regenerator.

2.2.7.5 Process Models

Models designed to predict the performance of an industrial process require information of many kinds: the properties and concentrations of the reacting species in the reaction medium, the relevant reaction rate coefficients and stoichiometries as well as a mechanistic description of the fluid flow in the reactor in which the process is carried out. In the case of catalytic cracking the modelling process is complicated by the fact that the feedstock contains thousands of chemical compounds distributed among the various classes—paraffins, naphthenes, aromatics etc. all capable of reacting with the catalyst at different rates and to varying degrees. Historically the approach to the problem of model formulation has been to divide the feed into "lumps" characterised according to one or more correlations based on physical and chemical properties of the hydrocarbon components (Astarita and Sandler 1984). In one such, the "n-d-M" method of Van Ness and Van Westen (1951), refractive index, density and molecular weight are used to identify carbon atoms in the various structures the lumps then being incorporated into a kinetic scheme to predict the performance of the unit. An early example of the method is provided by the work of the Mobil group (Weekman 1968; Voltz et al. 1971) who proposed) a three-lump system comprising unreacted gas oil, A_1, gasoline product, A_2, C_1 to C_4 gases and coke, A_3 combined in the following way:

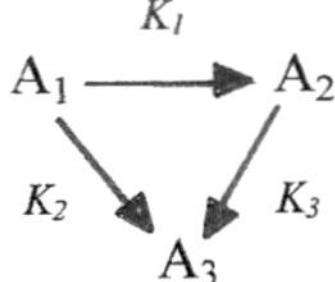

where the K terms represent rate coefficients. Reaction of the gas oil was assumed to be a second-order process while the cracking of the gasoline was assumed to be of the first order. Model equations describing the system for a vapour-phase, plug flow isothermal reactor were developed and tested experimentally against a number of charge stocks, good agreement being found (Voltz et al. 1971). To cater for a wider variety of feedstocks Jacob et al. (1976) developed a 10-lump scheme incorporating coke and light ends, gasoline, light paraffins, heavy paraffins, light naphthenes, heavy naphthenes, light aromatics, light aromatics with side chains and heavy aromatics with side chains. The model has formed the basis for subsequent more complex formulations such as the 19-lump model of Pitault et al. (1994) and the 21-lump model described by Chang et al. (2012). Individual lumps are identified within the boiling ranges obtained by fractionation of the crude, the composition of paraffins (P), naphthenes (N) and aromatics (A) being found from correlations such as that of Riazi (2005):

$$\%X_P \text{ or } \%X_N \text{ or } \%X_A = a + bR_i + cVGC$$

where the X terms represent molar compositions, R_i is the refractive index and VGC is a function of viscosity. The parameters a, b and c vary according to molecular type and boiling range. An additional term accounting for specific gravity can also be included.

Gupta et al. (2007) used a similar approach dividing the feed into 50 lumps, seven of which were the pure components from C_1 to C_5, the remaining 43 being pseudo-components identified according to a complex combination of boiling point and specific gravity. They also modelled the riser as a vertical tube comprising a number of equal-sized two-phase flow compartments in which each phase is well mixed and free from heat- and mass-transfer resistances.

Gao et al. (1999) developed a 3-D two-phase-flow reaction model for FCC risers combining two-phase turbulent flow with 13-lump kinetics. The model, based on the Eulerian two-fluid approach, illustrated the complexity of the feed injection zone at the base of the riser where flow fields, particle concentration, temperature and yield distributions showed significant inhomogeneities in the axial, radial and circumferencial directions. Nevertheless good agreement was found between model predictions and data from a commercial riser reactor.

An alternative approach has been to explore mechanistic models that track the chemical intermediates occurring in the FCC process using transition-state theory to quantify reaction-rate constants for adsorption and desorption at the catalyst surface. Froment has carried out pioneering work in this area (Feng et al. 1993; Froment 2005).

Chang et al. (2012) have reviewed the development since 1985 of so-called "unit-level" models that apply to an entire FCC unit i.e. riser reactor, stripper, regenerator, feed vaporiser, valves and cyclones, each item being described by a sub-model. Details of the sub-models have been given by Han et al. (2004). For a detailed description of the methodology using the Aspen HYSYS Petroleum Refining FCC model the reader is referred to the work of Chang et al. (2012).

References

Alizadeh M, Mostoufi N, Pourmahdian S (2004) Modeling of fluidized-bed reactor of ethylene polymerization. Chem Eng J 97:27–35

Astarita G, Sandler SI (1984) Kinetic and thermodynamic lumping of multicomponent mixtures. Elsevier, Amsterdam

Baker MJ, Couves JW, Griffin KG, Johnston P, McNicol JC, Salem GF (2003) Process for making a catalyst. US Patent 7053024

Bergna HE (1988) US Patent 4,769,477

Berruti F, Kalogerakis N (1989) Modelling the internal flow structure of circulating fluidized beds. Can J Chem Eng 67:1010–1014

Blum PR, Nicholas ML (1982) US Patent 4,317,778

Boland D, Geldart D (1971) Electrostatic charging in gas-fluidized beds. Powder Tech 5:289–297

Bolthrunis CO, Silverman RW, Ferrani DC (2004) Rocky road to commercialization: breakthroughs and challenges in the commercialization of fluidized-bed reactors. In: Fluidization XI. Engineering Conferences International, New York, pp 547–554

Brazdil JF (2005) Acrylonitrile, Ullman's Encyclopedia of industrial chemistry. Weinheim-Wiley-VCH, London

Burdett I D, Eisinger RS, Cai P, Lee KH (2001) Gas-phase fluidization technology for production of polyolefins. In: Fluidization X. United Engineering Foundation, New York, pp 39–52

Centi G (1993) Vanadyl pyrophosphate—a critical overview. Catal Today 16:5–26

Centi G, Fornasari G, Trifiro F (1985) n-butane oxidation to maleic anhydride on vanadium phosphorus oxides: kinetic analysis with tubular-flow stacked-pellet reactor. Ind Eng Chem Proc Des Dev 24:32

Chang AF, Pashikanti K, Liu YA (2012) Refinery engineering: integrated process modelling and optimization. Wiley-VCH, Weinheim

Chen Y-M (2003) Applications of fluidized catalytic cracking. In: Chapter 14 in handbook of fluidization and fluid-particle systems. Marcel Dekker, New York

Chen JQ, Bozzano A, Glover B, Fuglerud T, Kvisle S (2005) Recent advancements in ethylene and propylenre production using the UOP/Hydro MTO process. Catal Today 106:103–107

Chinh J-C, Filippelli MCH, Newton D, Power MB (1998) US Patent 5,733,510

Chiusoli GP, Maitlis PM (2008) Metal catalysis in industrial organic processes. RSC Publishing, Cambridge

Choi K-Y, Ray WH (1985) The dynamic behaviour of fluidized bed reactors for solid catalysed gas phase olefin polymerization. Chem Eng Sci 40:2261–2279

Contractor RM (1999) DuPont's CFB technology for maleic anhydride. Chem Eng Sci 54:5627–5632

Contractor RM, Sleight AW (1987) Maleic anhydride from C-4 feedstocks using fluidized-bed reactors. Catal Today 1:587–607

Cui HP, Mostoufi N, Chaouki J (2000) Characterization of dynamic gas-solid distribution in fluidized beds. Chem Eng J 79:135–143

Davidson JF, Harrison D (1963) Fluidised particles. Cambridge University Press, Cambridge

Dente M, Pierucci S, Tronconi E, Cecchini M, Ghelfi F (2003) Selective oxidation of n-butane to maleic anhydride in fluid-bed reactors: detailed kinetic investigation and reactor modelling. Chem Eng Sci 58:643–648

Dimian AC, Bildeak CS (2008) Acrylonitrile by propene ammoxidation. In: Chemical process design: computer-aided case studies, pp 313–338

Dry ME (1996) Practical and theoretical aspects of the catalytic Fischer-Tropsch process. Appl Catal A 138:319–344

Duvenhage DJ, Shingles T (2002) Synthol reactor technology development. Catal Today 71:301–305

Fakeeha AH, Solimam MA, Ibrahim AA (2000) Modelling of a circulating fluidized-bed for ammoxidation of propane to acrylonitrile. Chem Eng Proc 39:161–170

Feng W, Vynckier E, Froment GF (1993) Single event kinetics of catalytic cracking. Ind Eng Chem Res 32:2997

Fernandez FAN, L:ona LMF (2004) Multizone circulating reactor modelling for gas-solid polymerization: 1 reactor modelling. J Appl Polym Sci 93(3):1042–1052

Fernandez FAN, Lona LMF (2001) Heterogeneous modelling for fluidized-bed polymerization reactor. Chem Eng Sci 56:963–969

Fischer D, Frank H, Lux M, Hingman R, Schweier G (2000) US Patent 6,022,837

Froment GF (2005) Single event kinetic modelling of complex catalytic processes. Catal Rev Sci Eng 47:83–124

Fulks BD, Sawin SP, Aikman CD, Jenkins JM (1989) US Patent 4,876,320

Funk GA, Myers D, Vora B (2013) A Different Game Plan. Hydrocarbon engineering, December

Gao J, Xu C, Lin S, Yang G, Guo Y (1999) Advanced model for turbulent gas-solid flow and reaction in FCC risers. AIChE J 45:1095–1113

Gary JH, Handwerk GE (2001) Petroleum refining: technology and economics, 4th edn. Marcel Dekker, New York

Gayubo AG, Aguayo AT, Alonso A, Atutxa A, Bilbao J (2005) Reaction scheme and kinetic modelling for the MTO process over SAPO-18 catalyst. Catal Today 106:112–117

Glaeser LC, Brazdil JF, Toft MA (1989) US Patent 4,837,233

Goldbig KG, Werther J (1997) Selective synthesis of maleic anhydride by spatial separation of n-butane oxidation and catalytic reoxidation. Chem Eng Sci 52:583–595

Goode MG, Hasenberg DM, McNeil TJ, Spriggs TE (1989) US Patent 4,803,251

Grasselli RK (1999) Advances and future trends in selective oxidation and ammoxidation catalysts. Catal Today 49:141–153

Gupta RK, Kumar V, Srivastava VK (2007) A new generic approach for the modelling of fluid catalytic cracking riser reactor. Chem Eng Sci 62:4510–4528

Han IS, Riggs JB, Chung CB (2004) Modelling and optimization of a fluidized catalytic cracking process under full and partial combustion modes. Chem Eng Proc 43:1063–1084

Hanna TA (2004) The role of bismuth in the Sohio process. Coord Chem Revs 248:429–440

Harandi MN (1993) US Pstent 5,177,279

Hartge EU, Poggioda M, Reimers C, Schweir D, Gruhn G, Werther J (2006) Flowsheet simulation of solids processes. KONA 24:146–158

Hendrickson G (2006) Electrostatics and gas-phase fluidized-bed polymerization wall sheeting. Chem Eng Sci 61:1041–1064

Ibrehema AS, Hussaina MA, Ghasemb NM (2009) Modified mathematical model for gas-phase olefin polymerization in fluidized-bed catalytic reactor. Chem Eng J 149:353–362

Idol JD (1959) US Patent 2,904,580

Jacob SM, Gross B, Voltz SE, Weekman VW (1976) AIChE J 22:701–713

Jazayeri B (2003) Applications for chemical production and processing. In: Yang W-C (ed) Chapter 16 in handbook of fluidization and fluid-particle systems. Marcel Dekker, New York

Jiang P, Wei F, Fan L-S (2003) General approaches to reactor design. In: Yang W-C (ed) Chapter 12 in handbook of fluidization and fluid-particle systems. Marcel Dekker, New York

Kaarsholm M, Rafii B, Joensen F, Cenni R, Chaouki J, Patience GS (2010) Kinetic modelling of methanol-to-olefin reaction over ZSM-5 in fluid bed. Ind Eng Chem Res 49:29–38

Kaminski W (1998) Highly active metallocene catalysts for olefin polymerization. J Chem Soc Dalton Trans 1413–1418

Karri SBR, Werther J (2003) Gas distributor and plenum design in fluidized beds. In: Chapter 6 in handbook of fluidization and fluid-particle systems. Marcel Dekker, New York

Kiashemshaki A, Mostoufi N, Sotudeh-Gharebagh R (2006) Twi-phase modelling of a gas-phase polyethylene fluidized-bed reactor. Chem Eng Technol 61:3997–4006

Knowlton TM (2003) Cyclone separators. In: Chapter 22 in handbook of fluidization and flui-particle systems. Marcel Dekker, New York

Kunii D, Levenspiel O (1969) Fluidization engineering. Wiley, New York

Kunii D, Levenspiel O (1991) Fluidization engineering, 2nd edn. Butterworth-Heinemann, Boston

Mars P, van Krevelen DW (1954) Chem Eng Sci Suppl 3: 41

McAuley KB, Talbot JP, Harris TJ (1994) A comparison of two-phase and well-mixed models for fluidized bed polyethylene reactors. Chem Eng Sci 49:2035

McAuley KB, Macdonald DA, McLellan PJ (1995) Effects of operating conditions on stability of gas-phase polyethylene reactors. AIChE J 41:868–879

Mills PL, Randall HT, McCracken JS (1999) Redox kinetics of $VOPO_4$ with butane and oxygenusing the TAP reactor system. Chem Eng Sci 54:3709–3722

Minchiner A (2014) Made in China. Chem Engineer 872:42–45

Mostoufi N, Cui H, Chaouki J (2001) A comparison of two- and single-phase models for fluidized-bed reactors. Ind Eng Chem Res 40:5526–5532

Moughrabiah WO, Grace JR, Bi XT (2012) Electrostatics in gas-solid fluidized beds for different particle properties. Chem Eng Sci 75:198–208

Park T-Y, Froment GF (2004) Analysis of fundamental reaction rates in the methanol-to-olefin process on ZSM-5 as a basis for reactor design and operation. Ind Eng Chem Res 43:682–689

Patience GS (1990) Hydrodynamics and reactor modelling. PhD Dissertation, Ecole Polytechnique de Montreal

Patience GS, Bockrath RE (2010) Butane oxidation process development in a circulating fluidized bed. Appl Cat A 376:4–12

Patience GS, Bockrath RE, Sullivan JD, Horowitz HS (2007) Pressure calcination of VPO catalyst. Ind Eng Chem Res 46:4374–4381

Pitault I, Nevicato D, Forrissier M, Bernaed J-R (1994) Kinetic model besed on a molecular description for catalytic cracking of vacuum gas oil. Chem Eng Sci 49:4249–4262

Puettemann A, Hartge EU, Werther J (2012a) Application of flowsheet simulation concept to fluidized-bed reactor modelling. Part I: development of a fluidized-bed reactor model. Chem Eng Proc 60:86–95

Puettemann A, Hartge EU, Werther J (2012b) Application of flowsheet simulation concept to fluidized-bed reactor modelling. Part II: application to the selective oxidation of n-butane to maleic anhydride in a riser/regenerator system. Chem Eng Proc 57–58:86–95

Pugsley T, Patience GS, Berruti F, Chaouki J (1992) Modelling the catalytic oxidation of n-butane to maleic anhydride in a circulating fluidized-bed reactor. Ind Eng Chem Res 31:2652–2660

Riazi MR (2005) Characterization and properties of petroleum frsctions. ASTM, Conshohocken, PA

Roy S, Dudukovic MP, Mills PL (2000) A two-phase compartment model of the selective oxidation of n-butane in a circulating fluidized-bed reactor. Catal Today 61:73–85

Sadeghbeigi R (2012) Fluid catalytic cracking handbook, 3rd edn. Elsevier, New York

Sax NI (1975) Dangerous properties of industrial materials. Van Nostrand Reinhold, New York

Secchi AR, Neumann GA, Gambetta R (2013) Gas fluidized bed polymerization. In: Passos ML, Barrozo MAS, Mujumdar AS (eds) Chapter 2 in fluidization engineering: practice. Laval, Canada

Shamiri A, Hussain MA, Mjalli FS, Mostoufi N (2011) Dynamic modelling of gas-phase propylene homopolymerization in fluidized-bed reactors. Chem Eng Sci 66:1189–1199

Sookai S, Langanhoven PL, Shingles T (2001) Scale-up and commercial reactor fluidization-related experience with Synthol gas-to-liquid fuel dense phase fluidized-bed reactors. In: Fluidization X. United Engineering Foundation, New York, pp 621–626

Thompson LM, Bi H, Grace JR (1999) A generalised bubbling turbulent fluidized-bed reactor model. Chem Eng Sci 54:2175–2185

Van Ness K, Van Westen HA (1951) Aspects of the constitution of mineral oils. Elsevier, New York

Venuto PB, Habib TE (1979) Fluid catalytic cracking with zeolite catalysts. Marcel Dekker, New York

Vollert J, Werther J (1994) Mass transfer and reaction behaviour of a circulating fluidized-bed reactor. Chem Eng Technol 17:201–209

Voltz SE, Nace DM, Weekman VW (1971) Application of a kinetic model for catalytic cracking. Ind Eng Chem Proc Des Dev 10(4):530–541

Weekman VW (1968) A model for fluidized catalytic cracking. Ind Eng Chem Proc Des Dev 7:90

Wong R, Pugsley T, Berruti F (1992) Modelling the axial voidage profile and flow structure in the riser of a circulating fluidized bed. Chem Eng Sci 47:2301–2306

Xie T, McAuley KB, Hsu CC, Bacon DW (1994) Gas-phase ethylene polymerization: production processes, polymer properties and reactor modelling. Ind Eng Chem Res 33:449–479

Yamamoto R, Uetake S, Ohtani Y, Kikuchi Y, Doi K (1998) US Patent 5,753,191

Yates JG (1983) Fundamentals of fluidized-bed chemical processes. Butterworths, London

Zhou H, Wang Y, Wei F, Wang D, Wang Z (2008) Kinetics of the reactions of the light alkenes over SAPO-34. Appl Catal 348:135–141

Chapter 3
Non-catalytic Processes, Combustion, Gasification and Chemical Looping

Abstract In the processes to be considered here the advantageous features of fluidized beds noted in the previous Chapter are again in evidence. The high degree of solids mixing in multi-component systems such as are used in the chloride process for titanium dioxide and the consequent isothermal nature of the reacting mixture are highly desirable features and are exploited to the full in reactor design. The ability to transfer fluidized solids between reactors is exploited in the treatment processes of uranium compounds leading to the production of uranium dioxide for use in nuclear reactors. These are described along with processes for the production of hydrogen chloride and ultra-pure silicon while fluid coking and sulphide ore roasting are touched on briefly. The fluidized-bed combustion of coal is treated in detail with sections on plant developments, combustion mechanisms, desulfurization and sulfation models. Coal gasification is also considered and the chapter ends with an extended section on the relatively new technique of chemical looping.

3.1 Titanium Dioxide

Titanium is the ninth most abundant element in the Earth's crustal rocks, its two most important minerals being ilmenite ($FeTiO_3$) containing 40–70 % TiO_2 and rutile (TiO_2) which is about 95 % TiO_2. Rutile is the thermodynamically stable form at all temperatures. The two main processes for the production of titanium dioxide from these ores are (i) the wet sulphate process in which ground ilmenite is digested with sulphuric acid to produce a solution of the sulphates of titanium and iron from which the $FeSO_4$ is crystallized out leaving the $Ti(SO_4)_2$ to be calcined to TiO_2, and (ii) the chloride process using fluidized-bed reactors. The chloride process is continuous and the product normally is superior in colour to that from the sulphate process. Titanium dioxide is the most important white pigment, 80 % of the current world production of some 4.6 Mt/y being used in the paint, paper and plastics industries the balance being made up of applications in printing inks, fibres, cosmetics, foodstuffs, glass, electrical components and catalysts. In order to appear white a pigment must have minimal optical absorption at visible wavelengths and

J.G. Yates and P. Lettieri, *Fluidized-Bed Reactors: Processes and Operating Conditions*, Particle Technology Series 26,
DOI 10.1007/978-3-319-39593-7_3

Table 3.1 Refractive indices of some pigments and other materials (Greenwood and Earnshaw 1997)

Substance	Refractive Index	Substance	Refractive Index	Substance	Refractive Index
NaCl	1.54	$BaSO_4$	1.64–1.65	Diamond	2.42
$CaCO_3$	1.53–1.68	ZnO	2.0	TiO_2 (anatase)	2.49–2.55
SiO_2	1.54–1.56	ZnS	2.36–2.38	TiO_2 (rutile)	2.61–2.90

for TiO_2 this requires a high degree of chemical purity, in particular transition-metal impurities must be eliminated as far as possible. The use of TiO_2 as a constituent of sunscreen formulations depends on the high value of its refractive index in the visible region of the spectrum (Table 3.1) and its strong UV light absorbing properties.

Titanium dioxide is used in the production of self-cleaning window glass first introduced by the Pilkington Glass company in 2001. The windows are coated with a thin, 20–40 μm, layer of TiO_2 formed by chemical vapour deposition. The coating acts in two stages: first a photocatalytic stage in which sunlight (UV) breaks down deposits of organic dirt and causes the glass surface to become hydrophilic; this is followed by rain water washing the surface, spreading evenly over it as a sheet and removing the dirt. A number of other companies have since introduced a similar type of self-cleaning glass all employing surface coatings of titanium dioxide.

3.1.1 The Chloride Process

Here the TiO_2 content of the ore is frequently increased to over 90 % by beneficiation using a leaching technique. The process then involves mixing the beneficiated material with a source of carbon such as coke and reacting the two with chlorine in a fluidized-bed reactor at 900–1300 °C and 1.5–3 bar pressure:

$$\mathrm{TiO_2 + 2Cl_2 + C = TiCl_4 + CO_2; \Delta H = -212.7\,kJ/mol}$$

The highly exothermic reaction is carried out in a vessel of refractory-lined carbon steel, the good solids mixing characteristics of the fluid bed enabling the ore and coke to be brought into intimate contact under reaction conditions. Reactor diameters are typically in the range 2–8 m with multiple chlorine jets in the distributor plate; settled bed depths are in the range 2–8 m with operating gas velocities of 0.15–0.45 m/s. The fluidizing gas is predominantly chlorine (ca. 70 %) diluted with inerts such as nitrogen. The particle size of the titanium-bearing material is in the range 70–800 μm and that of the coke 300–5000 μm (Glaeser and Spoon 1995) putting them in Group B of the Geldart classification (Luckos and den Hoed 2004). As well as $TiCl_4$ and the chlorides of other metals present in the feed

(Fe, V, Cr etc.) a number of additional products are formed under the reaction conditions. Notable amongst these are carbonyl sulphide (COS) and sulphur dioxide formed from any sulphur compounds present in the feed, and carbon monoxide via the endothermic Boudouard reaction:

$$CO_2 + C = 2CO; \Delta H = 172.5\,kJ/mol$$

The optimum bed temperature is determined by the composition of the titanium-bearing feed and for efficient operation a balance must be struck between the heat generated by the exothermic chlorination reactions and that lost by (a) heating the incoming cold reactants to bed temperature, (b) the endothermic production of CO, and (c) the overall losses by conduction, convection and radiation. In the corrosive environment prevailing in the reactor thermocouples deteriorate relatively rapidly making direct measurement of bed temperature by this means difficult and as a result indirect methods are often employed. One such method uses the ratio of the concentrations of CO_2 and CO in the effluent gases as a measure of the extent of reaction and hence of bed temperature (Carlson and Mitchell 1971) while another technique monitors the effluent concentrations of COS and SO_2, it having been established that a consistent relationship exists between this ratio and bed temperature even at low COS concentrations (Elkins 1997). Bed-temperature regulation is achieved by introducing into the reactor air or oxygen if the temperature falls or, if it rises, a cold, inert material such as liquid $TiCl_4$, a procedure known as "pourback" (Elkins 1997), the input rates being determined by the measured levels of CO_2, CO etc. in the off gases. Turnbaugh et al. (2007) describe a method for monitoring the ratios of the above components in the effluent stream using infrared absorption spectroscopy and linking the measured values directly to the feed rate of the heating/coolant material.

Products of the process are $TiCl_4$ and chlorides of the metallic impurities present in the ore. The effluent gases are cooled and low volatile chloride impurities (of e.g. iron, manganese, chromium) are separated by condensation. The $TiCl_4$ is condensed to a liquid and distilled to give a pure liquid m.p. −24 °C, b.p. 136 °C. Vanadium-containing impurities, minute quantities of which lead to discolouration of the resulting TiO_2, are converted to vanadium trichloride by use of reducing agents, the resulting VCl_3 then being separated from the $TiCl_4$ by distillation. The titanium tetrachloride is then burned with oxygen or oxygen-enriched air in a flame or plasma at 1400–2000 °C to give the dioxide and chlorine which is recycled:

$$TiCl_4 + O_2 = TiO_2 + Cl_2; \Delta H = -180.8\,kJ/mol$$

The high temperature of the plasma process leads to the exclusive formation of the rutile crystalline form of the product (Winkler 2003).

3.2 Uranium Processing

Naturally occurring uranium exists in the form of three isotopes, ^{238}U (99.27 %), ^{235}U (0.72 %) and ^{234}U (0.006 %). All undergo radioactive decay with half lives of 4.468×10^9 y, 7.038×10^8 y and 2.45×10^5 y respectively but only ^{235}U is fissionable by thermal-neutron absorption and thus capable of being used in nuclear reactors for the production of "atomic" energy. The metal itself is extracted from its ores and purified by a series of hydrometallurgical procedures involving roasting, leaching, precipitation and solvent extraction leading to a pure aqueous solution of uranyl nitrate $UO_2(NO_2)_2$. This solution or the molten hexahydrate is converted to the trioxide by being sprayed into a bed of trioxide particles fluidized with air at 300–400 °C:

$$UO_2(NO_2)_2 \rightarrow UO_3 + NO_2 + NO + O_2; \Delta H = 332\,\text{kJ/mol}$$

The bed particles grow in size as the nitrate decomposes and deposits fresh UO_3. The enlarged particles are withdrawn and transferred either batchwise or continuously to a second reactor where they are fluidized with a mixture of hydrogen and nitrogen at 650 °C to form the dioxide:

$$UO_3 + H_2 \rightarrow UO_2 + H_2O; \Delta H = -100\,\text{kJ/mol}$$

The great majority of the 465 electricity-generating nuclear reactors currently in operation worldwide are light-water moderated designs using a uranium dioxide fuel enriched up to 3 % in ^{235}U. The enrichment process is carried out with gaseous UF_6 in gas centrifuges or diffusers. Uranium hexafluoride is made in two stages starting with the dioxide produced as above. In the first stage the UO_2 is reacted with gaseous hydrogen fluoride in a fluidized bed at 550 °C to give UF_4:

$$UO_2 + 4HF \rightarrow UF_4 + 2H_2O; \Delta H = 180\,\text{kJ/mol}$$

followed by fluorination:

$$UF_4 + F_2 \rightarrow UF_6$$

After enrichment the hexafluoride is converted back to the dioxide by reaction with superheated steam to give the solid uranyl fluoride, UO_2F_2, which is then reduced to UO_2 with hydrogen in either a fluidized bed or a rotary kiln. The final stage of preparation of the fuel involves the production of pellets of UO_2 by cold pressing followed by sintering in hydrogen at 1600–1700 °C.

3.3 Hydrogen Chloride

In the late 1960s the Israeli company Arad Chemicals started a process aimed at producing 250 kT/y of phosphoric acid. This was to be made from phosphate rock and hydrochloric acid the latter being produced by the hydrolysis of magnesium chloride at 900 °C in two fluidized -bed reactors operated in parallel:

$$MgCl_2 + H_2O = 2HCl + MgO$$

Each reactor was some 10 m in diameter and 12 m tall constructed from steel, lined with firebrick, fitted with external cyclones and filled to a depth of about 5 m with sand fluidized with air. The reaction temperature was achieved by burning crude oil in the beds whereupon a brine solution containing the magnesium chloride was sprayed in, the HCl product passing out overhead via the cyclones to the phosphate treatment unit. The plant was situated in the Negev desert close to the Dead Sea Works which were the source of the brine solution, an otherwise useless end-product of the Dead Sea salt extraction processes. Under the high temperature conditions in the reactors the magnesium oxide formed in the hydrolysis reacted with the sand particles of the bed to form magnesium silicate:

$$MgO + SiO_2 = MgSiO_3$$

causing the bed particles to grow to some 5 mm in diameter and thereby change the flow behaviour of the beds from Group B to Group D. Bed particles were removed periodically and replaced with fresh sand so maintaining a more or less constant solids inventory. The beds were characterised by a high degree of solids turbulence with gas bubbles some over 1 m in diameter bursting at the surface. From the very start the reactors suffered from operating problems one of which was the difficulty of maintaining steady-state operation at the design capacity. Introducing the brine solution caused the bed temperature to fall but increasing the oil flow to compensate resulted in much of it by-passing in the form of large bubbles and burning overhead in the freeboard space so threatening the integrity of the structure. The process only ever operated at half design capacity and when the price of crude oil rose dramatically following the Yom Kippur war of 1973 the process became totally uneconomic and was shut down.

3.4 Ultra-Pure Silicon

Worldwide demand for this material from the semiconductor and photovoltaic industries has increased some six-fold over the last decade to a current (2010) figure of around 120 kt/a (Sabino et al. 2013). Permitted impurity levels for electronic-grade silicon are of the order of parts per billion (99.999999999 % Si) while those for the solar-grade material are somewhat higher at parts per million

(99.99999 % Si). The main starting material for the production of both grades is metallurgical-grade silicon, MGS, (98.5–99 % Si) produced from high-purity silica, SiO_2, in electric-arc furnaces for use in steel making and aluminium processing. At the present time three commercial processes are being applied worldwide (Caccaroli and Lohne 2011). In the widely-used Siemens process gaseous trichlorosilane, $SiHCl_3$, is cracked at 1100 °C on electrically-heated silicon rods mounted inside bell-shaped reactors (Fig. 3.1):

$$SiHCl_3 \rightarrow Si + SiCl_4 + 2HCl$$

The starting material is produced by the hydrochlorination of MGS in a fluidized-bed reactor:

$$Si + 3HCl \rightarrow SiHCl_3 + H_2; \Delta H = -496.2\,kJ/mol$$

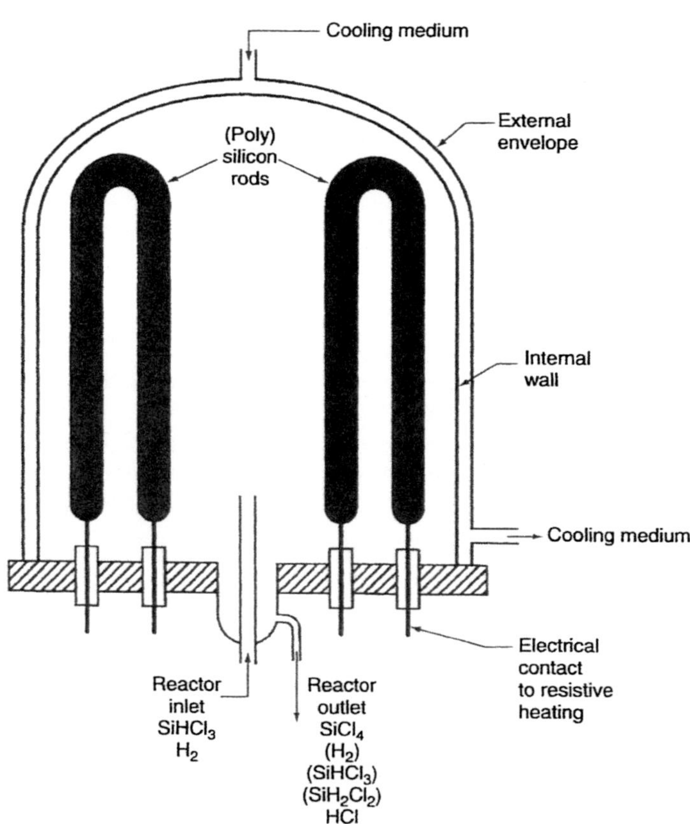

Fig. 3.1 Schematic representation of a traditional Siemans CVD reactor (Caccaroli and Lohne 2011)

In addition to $SiHCl_3$ other chlorosilanes are produced according to:

$$Si + (4 - n)HCl \rightarrow SiH_nCl_{4-n} + (2 - n)H_2$$

and in order to maximise the yield of trichlorosilane the reaction temperature must be controlled within the range 300–350 °C which, given the highly exothermic nature of the reactions, favours the use of a fluidized bed. A schematic of one version of the process is shown in Fig. 3.2. Pure $SiHCl_3$ is obtained from the reaction products by fractional distillation.

The Siemens process suffers from a number of disadvantages (Caccaroli and Lohne 2011): energy consumption is high, 90 % of the power input to the silicon rods being lost to the cold walls of the reaction vessel; hot-spot formation and filament burn-out may occur; the large amounts of by-products need to be treated or recycled. To overcome some of these problems a process based on the use of cheap,

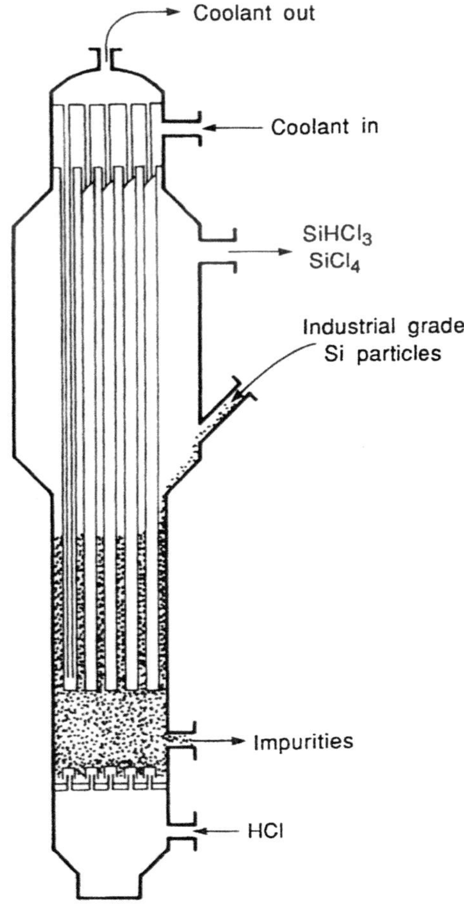

Fig. 3.2 Fluidized-bed silicon-hydrochlorination reactor (Kunii and Levenspiel 1991)

commercially available tetrachlorosilane, $SiCl_4$, was developed in the 1970s by Union Carbide (now Dow) and Komatsu Electronic Materials.

The starting point is the hydrogenation of $SiCl_4$ in a fluidized bed of MGS particles to give $SiHCl_3$:

$$3SiCl_4 + 2H_2 + Si \rightarrow 4SiHCl_3$$

The trichlorosilane product is then converted to silane, SiH_4, in fixed-bed reactors of quarternary ammonium ion-exchange resins:

$$2SiHCl_3 \rightarrow SiH_2Cl_2 + SiCl_4$$
$$3SiH_2Cl_2 \rightarrow SiH_4 + 2SiHCl_3$$

The silane is separated by distillation then pyrolysed at 800 °C in modified Siemans reactors to give silicon:

$$SiH_4 \rightarrow Si + 2H_2$$

The third process for the production of ultra-pure silicon is that originated by the Ethyl Corporation and later developed by MEMC Electronic Materials (now SunEdison) (Jazayeri 2003). The starting material for this process is an alkaline fluorosilicate such as $NaSiF_6$ a by-product of the fertilizer industry. Silicon tetrafluoride is obtained from this by thermal sublimation then hydrogenated to silane by reaction with lithium aluminium hydride:

$$SiF_4 + LiAlH_4 \rightarrow SiH_4 + LiAlF_4$$

The silane is then fed with hydrogen to a bubbling fluidized bed of high-purity silicon seed particles where it decomposes and deposits pure silicon onto the seed material. However homogeneous decomposition of silane also occurs in the void spaces of the bed leading to the formation of silicon dust particles below 10 μm in size which are elutriated from the reactor and represent an economic loss. A schematic of the reactor is shown in Fig. 3.3.

Reported operating conditions are 600–800 °C, 1 bar pressure and gas velocity of 1.2–3.5 times minimum fluidization; particle sizes are in the range 700–1100 μm (Jazayeri, 2003). The complex chemistry and hydrodynamics of the process have recently been explored by Sabino et al. (2013).

3.5 Fluid Coking

This process was developed by Exxon (now ExxonMobil) in the 1950s as a way of converting residual petroleum fractions into coke, oil and gas. The operating principle is analogous to the FCC process. Coke particles 150–200 μm in diameter

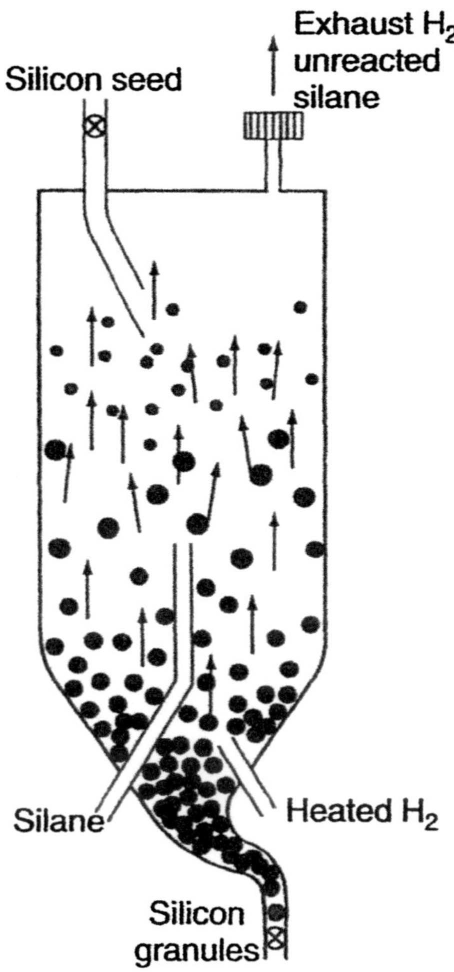

Fig. 3.3 Schematic of a fluidized-bed reactor for the production of ultra-pure silicon (Caccaroli and Lohne 2011)

are circulated between a cracking reactor at 500–600 °C where they are fluidized by steam and a heater fluidized by air in which a portion of the coke is burned off raising its temperature to 600–700 °C. The heated coke is returned to the reactor where the residuum feed is injected and cracked depositing a thin layer of coke on the particles which thereby acquire a roughly spherical shape. Cooled coke particles leave the reactor via a steam stripper where adsorbed hydrocarbons are removed while gaseous reaction products pass overhead via cyclones into a scrubber where they are quenched and fractionated. Up to 2003 thirteen commercial units had been built with capacities of up to 100,000 barrels/day (16,000 m^3/day) (Jazayeri 2003).

Flexi-Coking is an extension of fluid coking in which an additional coke-gasification reactor is incorporated to generate refinery fuel (Jazayeri 2003).

3.6 Sulfide Ore Roasting

The fluidized-bed technique of roasting sulfide ores such as pyrite, FeS_2, was developed in the 1950s by Dorr-Oliver in the USA, Badische Analine (BASF) in Germany and Sumitomo in Japan. The exothermic reaction is carried out by fluidizing ore particles under 10 mm in size with air in large diameter bubbling beds at 650–1100 °C:

$$FeS_2 + O_2 \rightarrow Fe_2O_3/Fe_3O_4 + SO_2$$

the SO_2 being used to manufacture sulfuric acid. Large numbers of such plants have been built around the world by Dorr-Oliver and Lurgi, the reported largest being that at Pasminco EZ in Hobart, Tasmania roasting 900 t/day of zinc sulfide ore in a bed of 16 m diameter (Dry and Beeby 1997). In the 1980s a circulating fluidized-bed roaster was developed by Lurgi with the object of roasting gold-bearing pyrite to liberate the gold content. The resulting unit, located at Gidgie in Western Australia, was designed to treat 535 t/day of ore concentrate and is reported to have operated successfully (Dry and Beeby 1997).

3.7 Fluidized-Bed Combustion

Solid fuels ranging from coals to biomass may be burned in fluidized beds for the purpose of raising steam for process heating, power generation via the Rankine cycle or for combined heat and power systems. The normal mode of operation is to feed the fuel at concentrations of up to a few percent by weight into a bed of either an inert material such as silica sand or ash or an active material such as limestone fluidized by air at temperatures in the range 750–900 °C. The beds may be operated in the low-velocity bubbling mode or as high-velocity circulating beds; both atmospheric-pressure and elevated-pressure systems have been commercialised successfully. In the case of coal combustion a number of features make the process an attractive alternative to conventional combustion systems such as pulverized fuel (PF) burners and chain-grate stokers:

(i) The relatively low operating temperature is advantageous in several ways. Thus it prevents the fusion of coal ash leaving it soft and friable and so reducing its erosion effect on exposed heat-transfer surfaces; it limits the volatilization of corrosive alkali-metal salts, vanadium etc. present in the coal and also limits the formation of oxides of nitrogen, NO_x.

(ii) Bed-to-immersed surface heat-transfer coefficients (100–400 W/m^2K) are up to an order of magnitude greater than those in gas-to-surface heat exchangers resulting in a considerable reduction in unit costs compared to conventional systems. Circulating beds operate at the lower end of this scale while bubbling beds work at the upper end (Brereton 1997).

(iii) Incorporating batches of limestone or dolomite in the bed material enables the gaseous SO_2 formed from the sulphur content of the coal to be converted to calcium sulphate and so retained in solid form thus permitting the pollution-free combustion of high-sulphur coals.

(iv) Low value fuels such as lignites, oil shale, washery tailings and peat which are largely incombustible in conventional systems are readily burned in fluidized beds. The 250 MWe CFB plant operated since 1996 by EDF at Gardanne in France has burned a heavy-sulphur local coal, a French hard coal, imported coals from South Africa and petroleum coke (Leckner 1998). This ability to burn a variety of fuels in the same unit is a major advantage.

(v) Burning rates in excess of 2.5 MW/m^2 based on distributor area have been achieved, values in excess of those typical of chain-grate stokers and comparable to pulverised-fuel combustors (Skinner 1970).

(vi) The ability to burn fuels of high moisture content.

(vii) Bubbling-bed combustors do not require the expensive pulverising equipment of PF systems and unlike stoker-grate burners are insensitive to fuel size distribution (Brereton 1997).

3.7.1 Plant Developments

(a) Bubbling fluidized-bed combustors (BFBC)

Interest in fluidized-bed combustion of coal developed strongly in several countries in the 1960s. Chinese work in this period was described by Zhang (1980) while Elliot (1970) and Bishop (1970) described early developments in the UK and USA respectively. The original aims of the development programmes in the UK and USA were quite different (Elliot 1970), the British being concerned with potential reductions in the capital costs of power stations while the Americans placed more emphasis on pollution control. Both programmes however progressed along similar lines. With the discovery of natural gas deposits under the North Sea in the mid-1960s however interest in the subject declined in the UK although it continues to be exploited in a number of other European countries such as Finland, Sweden, Poland, France and Germany by companies such as Ahlstom, Lurgi and Studsvik as well as in China, Japan and the USA where Foster Wheeler and Metso Power are major contractors (Lee 1997; Johnsson 2007).

The majority of the early work was concerned with BFBC's in which process steam was generated via water-fed, horizontally mounted heat-exchanger tubes immersed in the bed. Figure 3.4a shows an example of such a system where coal is introduced both underbed through the air distributor and overbed onto the bed surface.

The relative merits of underbed and overbed feeding have been discussed by Castleman (1985) in relation to the TVA 20 MWth demonstration unit. Underbed feeding requires up to one pneumatic feed point of coal fines per 1–2 m^2 of bed cross-sectional area to achieve high combustion efficiency (96–98 %) while over-bed feeding requires a coarser coal top size if around 25 mm (Newby 2003). Typical excess air levels are 20–25 % with dense beds 1–1.5 m in depth, splash zones some 10 m above the bed surface and fluidizing-air velocities of 1.5–4 m/s (Newby 2003). At these gas velocities considerable elutriation occurs particularly of the smallest particles and to maintain combustion efficiency any carbon so removed from the bed either as fines or combined in the fly ash must be returned via cyclone collectors or burned in carbon burn-up cells (Basu 2006).

As mentioned above in early designs of BFBC units bed temperature was controlled by means of immersed tubular heat exchangers. These however were prone to severe erosion by the moving bed solids and subsequent designs abandoned in-bed coolers in favour of membrane tubes located in the bed walls. Many current designs burning low-value wastes or biomass dispense with in-bed cooling altogether and rely on the moisture and volatile content of the fuel to control bed temperature, heat being extracted downstream of the combustion chamber in the "back-pass", the convective heat-transfer section located downstream of the combustor (Johnsson 2007). A serious drawback of BFBC's is their limited ability to

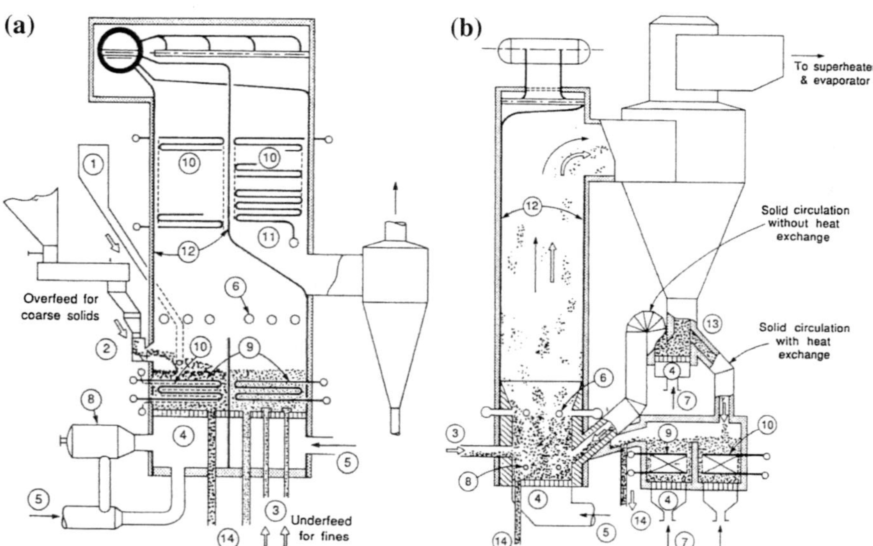

Fig. 3.4 Fluidized-bed combustors: **a** BBFB; **b** CFB. *1* Limestone Shute, *2* Spreader feeder, *3* Coal-limestone feeder, *4* Air distributor, *5* Primary air inlet, *6* Secondary air nozzle, *7* Fluidized air, *8* Hot gas generator, *9* Evaporator, *10* Superheater, *11* Economizer (water preheater), *12* Water wall, *13* Circulator, *14* Bed drain (Kunii and Levenspiel 1991)

operate under part load. In order to vary the load, i.e. the rate of steam generation, the rate of heat transfer to the heat exchanger surfaces must be varied. This can be achieved in a limited number of ways e.g. (i) if in-bed heat exchanger tubes are incorporated in the design reducing the bed height by lowering fluidizing gas velocity enables a proportion to be uncovered and so lower the heat transfer rate; (ii) designing the combustion chamber on a modular basis and slumping one or more modules in line with a reduced load requirement. These and other restricted techniques have been discussed by Brereton (1997), Basu (2006), Oka (2004), and Leckner et al. (2011).

Bubbling-bed combustors operating at atmospheric pressure are more suitable for small-capacity steam generators of up to 100 MWth or for combined heat and power systems although several large-capacity units have been built and operated successfully (Anderson 1997; Takahashi et al. 1995). Basu (2006) noted that at that time over 10,000 BFBC boilers were in use worldwide.

(b). Circulating fluidized-bed combustors (CFBC)

Circulating fluidized-bed combustors are widely used for power generation of up to 300 MW_e burning bituminous coal or lignite or co-firing with coal and biomass (Johnsson 2007). They are tall in comparison with BFBC units with height-to-diameter ratios of up to 10 (Fig. 3.5b). They are normally of square or rectangular cross-section with a tapered lower section and membrane water-wall sides. Primary air, 30–100 % of the total, is preheated and supplied through the distributor plate at the base at a pressure of 10–20 kPa; secondary air is introduced some distance above the base at a relatively low pressure of 5–15 kPa the total gas velocity being in the range 5–8 m/s with solids-to-gas mass ratios of between 3 and 30:1 (Brereton 1997). Bed solids are relatively coarse Group B materials although a proportion of finer material will be present as ash and unburned coal fines. Solids net fluxes range from 15 to 90 kg/m^2s (Davidson 2000). Operating conditions are significantly different from those typical of circulating fluidized-bed systems used in the process industries such as in fluidized-catalytic cracking; Table 3.2 highlights some of these differences.

Table 3.2 Comparison of typical operating conditions for FCC units and CFB boilers (Grace 1990)

Application	Boilers	Reactors
Regime	Turbulent to fast	Entrained flow
Geldart group	B to D	A, C or B
Mean particle diameter (μm)	200	70
Circulation rate (kg/m^2s)	<40	>
Solids residence time (s)	300–600	3–15
Gas velocity (m/s)	<6	<25
Temperature (°C)	750–900	400–500
Riser aspect ratio	<10	~20

In the riser section of a CFBC bed solids are in bubbling or turbulent flow in the base region passing via a transition zone into fast flow at higher points where a core-annulus structure exists with a dilute gas-solid mixture of single particles and clusters or strands flowing upwards in the core and denser clusters moving downwards at the walls (Fig. 3.5).

Davidson (2000) calculated the slip velocity, v_s, between gas and particles as follows:

$$v_s = u_G - u_S = \frac{U}{\varepsilon} - \frac{m}{\rho_S A(1-\varepsilon)} \tag{3.1}$$

where u_G and u_S are absolute velocities of gas and solids respectively, U is the superficial gas velocity, ε the mean void fraction, ρ_S the particle density, A the riser cross-sectional area and m the solids feed rate to the riser. Based on values of the solids volume fraction at various heights in a 0.4 m diameter riser determined by

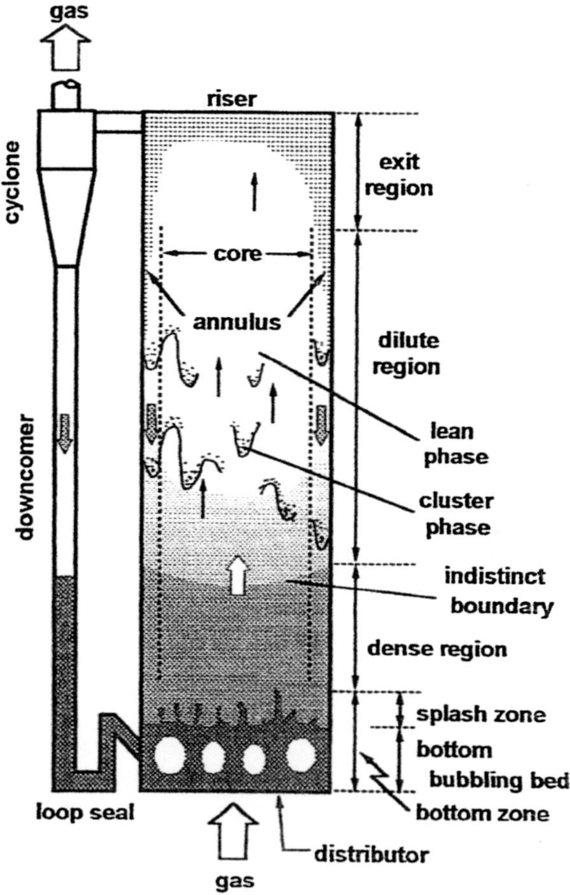

Fig. 3.5 Typical flow structure in a circulating fluidized bed (Horio1997)

Hartge et al. (1986) Davidson calculated v_s in the bubbling/turbulent zone and the upper fast region to be 4.8 and 2.4 m/s respectively, both greatly in excess of the terminal fall velocity of a single particle (0.3 m/s). He showed the value in the lower zone to be consistent with slugging bed behaviour (where the estimated slug velocity is 4.69 m/s) while in the upper region the slip velocity was consistent with the presence of particle clusters and streams of particles flowing downward at the walls.

Downward velocities in large units have been estimated at 2–8 m/s (Werther 2005) although Glicksman (1997) gives a value of 1 m/s while Rhodes et al. (1991) from high-speed video observations found values of 0.3–0.4 m/s with a region at 1 m/s; measured solids mass fluxes in a CFB boiler are shown in Fig. 3.6.

Interchange between core and annulus results in solids circulation patterns whose rate may be many times that of the external circulation rate (Horio and Morishita 1988). Further evidence for the zone structure of a CFBC is given by the study of the 12 MWth boiler at Chalmers University by Johnsson and Leckner (1995) (Fig. 3.7).

Pallares and Johnsson (2006) presented a semi-empirical model of a CFBC in which the unit was divided into six zones as shown in Fig. 3.8.

No allowance was made for the chemical or heat-transfer behaviour of the system, the emphasis being on fluid dynamics. The bottom bed was modelled as a bubbling bed characterised by emulsion and bubble phases with interphase exchange of gas (Johnsson et al. 1991). The freeboard was described in terms of the aforementioned core-annulus structure, particles being ejected into the zone from the lower bed by vigorously bursting bubbles. An important property of particles in the freeboard was shown to be their terminal fall velocity in an upward-flowing suspension, u_t, a value lower than that for an isolated particle; values of u_t were found from the model of Palchonok et. al. (1997).. Solids concentrations in the freeboard, which decay exponentially from the top of the bottom bed to the outlet, were estimated from a number of published correlations such as those of Johnsson and Leckner (1995) and Adanez et al. (1994). The pressure drop across the exit duct

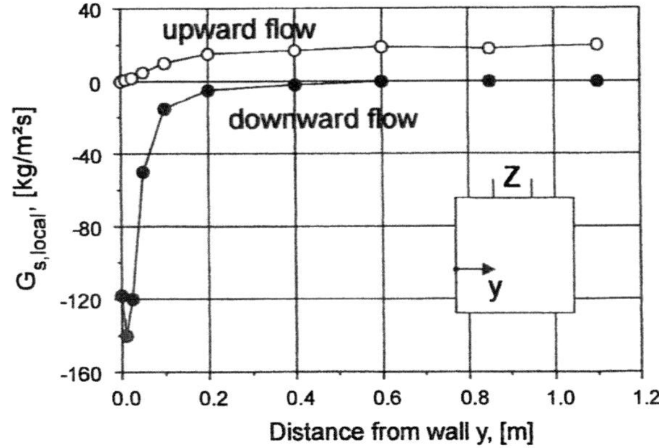

Fig. 3.6 Local solid-mass fluxes in a CFB boiler plotted against distance, y, from the wall for the Fensburg power plant; H = 28 m, x = 17.3 m, U = 6.3 m/s (Werther and Hirschberg 1997)

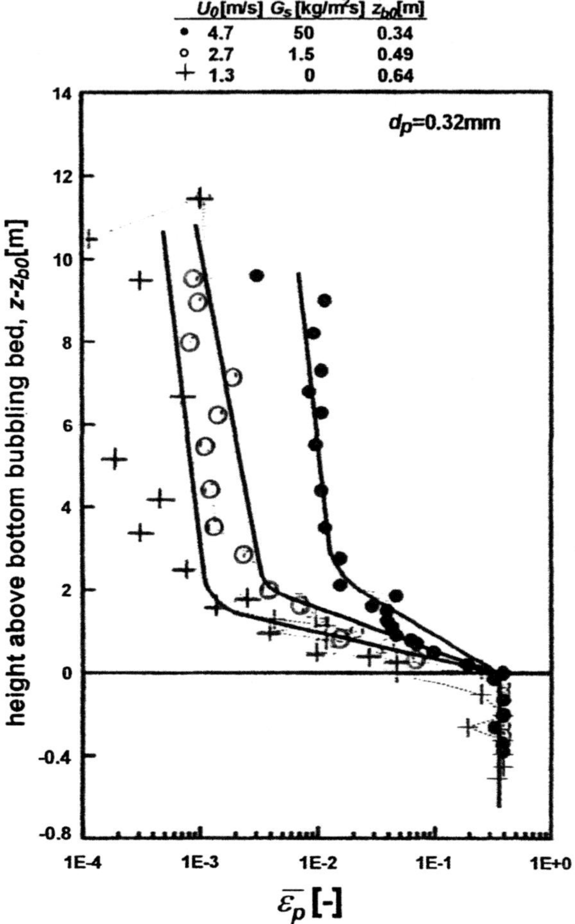

Fig. 3.7 Bed-density distribution in the bottom region of a 12 MW_{th} CFB boiler (Johnsson and Leckner 1995)

was determined from the model of Muschelknautz and Muschelknautz (1991) and that of Rhodes and Geldart (1987) which was also used to estimate the pressure drop across the cyclone. The downcomer and particle seal serves to return recycled particles to the riser and prevent gas bypassing into the cyclone.

The fluidized bed in the particle seal unit was modelled on the same basis as the lower section of the riser i.e. as a two-phase emulsion-bubble bed (Johnsson et al. 1991). The authors discuss the interactions among the six sub-units of the CFB and compare the resulting predicted solids concentration profiles with data from two commercial CFB combustors. Finally the solids concentrations and pressure distribution around the Chalmers CFB boiler were predicted and are as shown in Figs. 3.9 and 3.10.

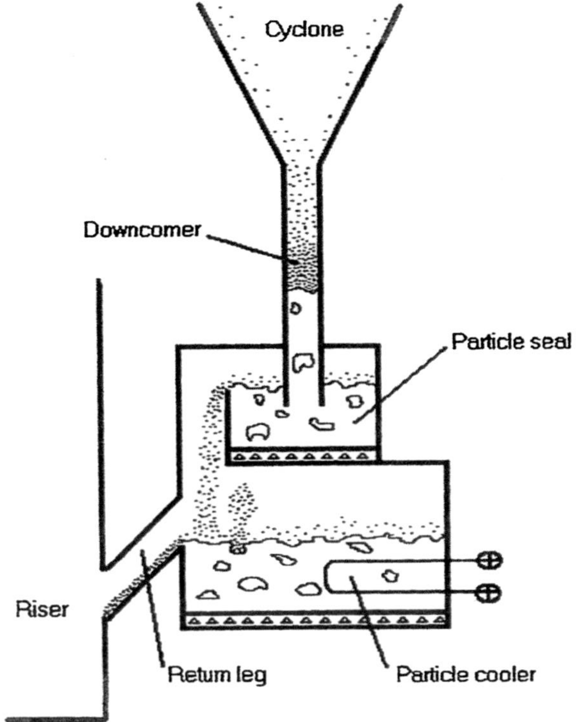

Fig. 3.8 Zone division in a CFB model (Pallares and Johnsson 2006)

The paper describes in detail the complexities of the model formulation and concludes that integration of the individual models into an overall CFB model requires that a population balance of the circulating solids is necessary for satisfactory predictions.

In the last decades a large number of purely hydrodynamic models have been proposed to describe the flow regimes in CFB systems. This work has been comprehensively reviewed by Horio (1997) and Sinclair (1997) while Louge (1997) has reviewed the experimental techniques developed to test the models. Despite these efforts a complete description of CFB hydrodynamics remains to be achieved although some progress in this direction has been made using computational fluid dynamics.

3.7.2 *Mechanism of Coal Combustion in Fluidized Beds*

The mechanism of the fluidized-bed combustion of coal or any other fuel such as biomass is a complex process involving the interplay of chemical reactions, heat and mass transfer, and bed hydrodynamics. Coal particles introduced into a bubbling air-fluidized bed at temperatures in the range 750–900 °C pass through a

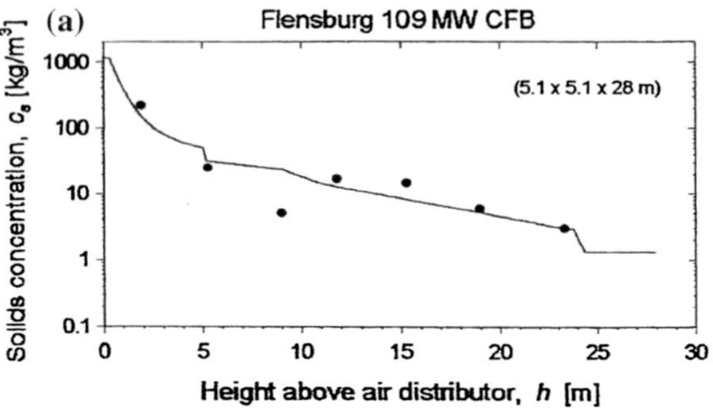

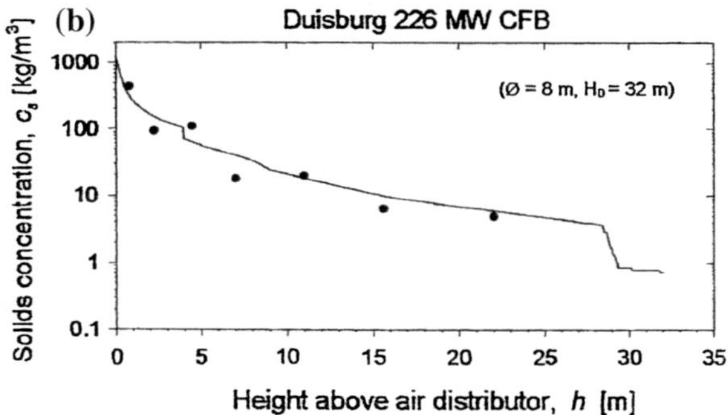

Fig. 3.9 A CFB return loop with independent particle cooler (Pallares and Johnsson 2006)

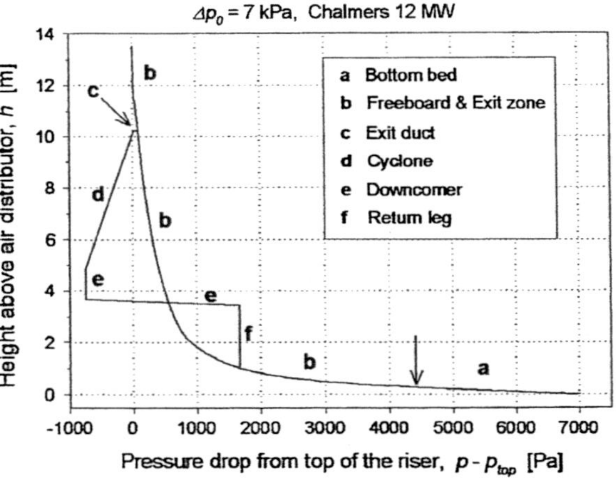

Fig. 3.10 Pressure diagram predicted by the overall CFB model for a standard run in the Chalmers CFB boiler (Pallares and Johnsson 2006)

number of stages the most important of which are devolatilization and combustion of the residual char. The time scales of these processes are mainly functions of bed temperature and the nature and size of the coal particles but in general devolatilization is an order of magnitude faster than char combustion. Based on a shrinking-core model La Nauze (1985) showed the devolatilization time, t_v, to be proportional to the square of the coal particle diameter and the fractional yield, x_v, to be given by:

$$\frac{t_v}{\tau} = 1 - 3(1 - x_v)^{2/3} + 2(1 - x_v) \tag{3.2}$$

where τ is the time for complete devolatilization. On evolution the volatile component, which can comprise up to 50 % of the heating value of the fuel, burns in a diffusion flame surrounding the particle (Agarwal 1986), the burnout time in seconds, T_v, being correlated for 12 different coals by the empirical expression:

$$T_v = ad_p^N \tag{3.3}$$

where d_p is the initial diameter of the coal particle in millimetres, N is a function of bed temperature, T_b, and coal type and varies from 0.32 to 1.8, and a is an empirical constant varying between 0.22 and 22 which is proportional to $T_b^{3.8}$.

A number of system models have been developed to describe volatiles combustion in bubbling fluidized beds (La Nauze 1985) among which is the so-called "plume" model of Park et al. (1980). The model is based on the assumption that since coal devolatilization is fast (1 s for a 200 μm particle) compared with both particle circulation and char combustion (60 s for a 300 μm particle) the particles entering at the distributor will devolatilize practically instantaneously and the volatiles will rise in plug flow as plumes from each entry point and burn in a diffusion flame with the fluidizing air at the plume boundary (Fig. 3.11).

This combustion can be completed either in the bed or, if the plume breaks through the bed surface, in the freeboard the latter circumstance causing a temperature rise in the space above the bed. The conversion of volatiles is governed by the dimensionless "plume group" $HD_r/(UL^2)$ where H is the bed height, D_r a radial

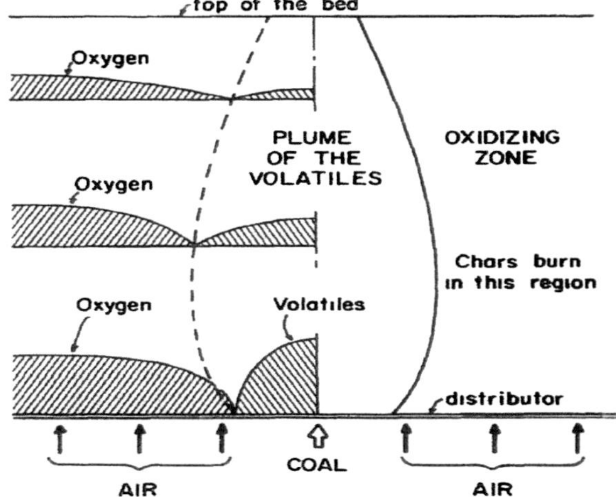

Fig. 3.11 Plume in the bed (Park et al. 1980)

dispersion coefficient and L the radius of a coal-feed module; for sufficiently large values of the group the volatiles burn completely in the bed.

A number of studies of char combustion in fluidized beds were carried out by the Cambridge University group in the 1970s and ‘80s and led to the publication of models describing the burn-out time of batches of materials. The early work of Avedesian and Davidson (1973) was based on the assumption that the combustion rate is controlled by the rate of diffusion of oxygen to the surface of the burning particle and that chemical reactions are so fast at the temperatures in question that they have no effect on the overall kinetics. This latter assumption was questioned by Basu et al. (1975) and led to the development by Ross and Davidson (1982) of a modified expression for the burn-out time that incorporated both diffusion and chemical effects. Here an overall rate constant K is defined as a combination of kinetic and diffusional terms:

$$\frac{1}{K} = \frac{\gamma}{k_c} + \frac{\alpha d_p}{ShD_G} \tag{3.4}$$

where k_C is a reaction rate coefficient, Sh is a Sherwood number varying from $2\varepsilon_{mf}$ (where ε_{mf} is the dense-phase voidage) for large particles to 2 for particles smaller than the bed material of size d_p, D_G is the molar diffusivity of oxygen; α and γ are constants whose values depend on the reactions assumed to be occurring at the char surface:

	α	γ
(i) $C + \frac{1}{2}O_2 \rightarrow CO$	1	2
(ii) $C + O_2 \rightarrow CO_2$	2	2
(iii) $C + CO_2 \rightarrow 2CO$	1	1

The resulting expression for the burn-out time, t_c, of a batch of char in a bed of cross-sectional area A, derived on the basis of the Davidson-Harrison model for a completely mixed emulsion phase, is:

$$t_c = \frac{m}{12C_0A\left[U-(U-U_{mf})e^{-X}\right]} + \frac{\rho_c d_i^2 \alpha}{96ShD_GC_0} + \frac{\rho_c d_i \gamma}{48k_cC_0} \tag{3.5}$$

Experimental results combined with theoretical calculations of Borghi et al. (1977) led Ross and Davidson to conclude that the combustion rate is controlled by a combination of the two effects. For large particles ($d_p > 3$ mm) the rate is controlled by diffusion of oxygen to the surface; in the case of small particles however, combustion is controlled mainly by the kinetics of the reaction $C + \frac{1}{2}O_2 \rightarrow CO$, the CO so formed then burning in the emulsion phase around the particles. In the latter case the heat of combustion of the CO is dissipated in the emulsion solids with the result that small carbon particles are likely to be at or near the bed temperature whereas large particles are generally hotter than the bed.

The combustion of a pseudo-char in a circulating fluidized-bed combustor was studied experimentally by Basu and Halder (1989) using a 102 mm diameter and 5.5 m high CFB made of stainless steel and operated at around 800 °C. The bed material was sand and the pseudo-char was composed of 5–9 mm diameter spherical particles of electrode carbon chosen instead of char itself to avoid the problem of fragmentation on heat-up. Burning rates of the carbon particles were measured over a range of gas velocities and oxygen concentrations and the results compared with an empirically-based theoretical model. In the model the mass, momentum and energy balance equations were given and solved numerically. The mass balance was expressed as:

$$\frac{dm}{dt} = -qA - k_a^{'}\left(U_c - U_p\right)A \tag{3.6}$$

where m is the mass of a single carbon particle (kg), q is the carbon burning rate (kg/m^{2}s), $k_a^{'}$ is an attrition rate constant (kg/m^3), U_c and U_p are the average upward velocities of bed solids and carbon particles respectively (m/s) and A is the surface area of a carbon particle (m^2). The second term on the right hand side accounts for attrition of the coarse particles caused by the slip between them and the upward-moving fine solids. Carbon burning rates were related to:

$$\frac{q}{\left(1 - q/m_d\right)^n} = R_c P^n \tag{3.7}$$

where q is the burning rate of carbon (kg/m^{2}s), m_d is the mass transfer limit or the maximum possible burning rate (kg/m^{2}s), n is the order of reaction, R_c is the reactivity of the fuel (kg/m^{2}s kPa^n) and P is the oxygen concentration (kPa). The ratio $\frac{q}{m_d}$ was defined as the fraction of mass-transfer control, X, and was found experimentally to lie between 0.55 and 0.65 showing combustion to be a combination of oxygen mass transfer and chemical kinetics. X was also found to decrease with decrease in carbon particle diameter indicating an approach towards kinetic control, the measured trend showing that for 5 mm diameter particles X would be less than 0.5. The experimental results and model predictions are shown in Fig. 3.12.

3.7.3 Desulfurization

During the combustion of coal the sulphur compounds present, thiols, thiophens, sulphides etc., are oxidized to sulphur dioxide which in the presence of the basic sorbent calcium oxide will react to form the solid calcium sulphate and so be

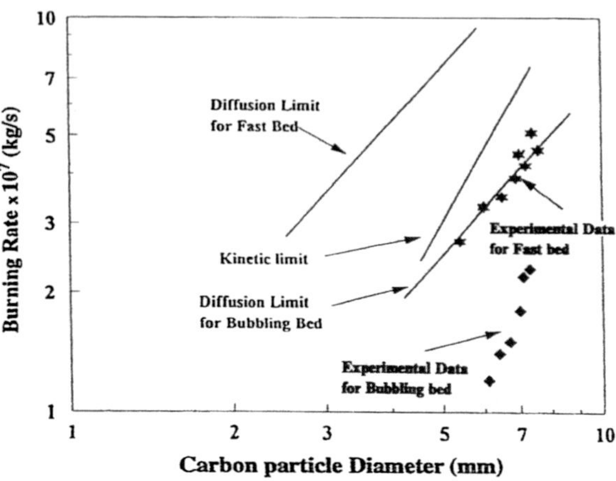

Fig. 3.12 Char combustion rates in a CFB boiler (Brereton 1997, after Basu and Halder 1989)

removed from the effluent gases. Calcium oxide may be formed in situ in a fluidized-bed combustor by the calcination of either limestone:

$$CaCO_3 \rightarrow CaO + CO_2; \Delta H = 183\,kJ/mol$$

or dolomite:

$$CaCO_3 \cdot MgCO_3 \rightarrow CaO \cdot MgO + 2\,CO_2$$

although in this case owing to the slow rate of reaction between MgO and SO_2 only the calcium component is sulphated under FBC conditions. The overall sulfation reaction of calcium oxide may be written as:

$$CaO + SO_2 + {}^1\!/_2\,O_2 \rightarrow CaSO_4;\ \Delta H = -471.9\,kJ/mol$$

although this does not describe the basic mechanism which may involve one or other of the following routes both of which are thermodynamically feasible under FBC conditions.

Route 1: $CaO + SO_2 \rightarrow CaSO_3$
$CaSO_3 + {}^1\!/_2\,O_2 \rightarrow CaSO_4$

Route 2: $SO_2 + {}^1\!/_2\,O_2 \rightarrow SO_3$
$CaO + SO_3 \rightarrow CaSO_4$

Dennis and Hayhurst (1990) however doubt whether either adequately describes the detailed mechanism. From an experimental study of the system they concluded that the reaction involves a complicated mechanism including:

$$CaO + SO_2 \leftrightarrow CaSO_3$$
$$4CaSO_3 \rightarrow CaS + 3CaSO_4$$
$$CaS + 2O_2 \rightarrow CaSO_4$$

together with the participation of other species $S_yO_x^{n-}$. They also concluded that absorption via SO_3 plays a negligible role.

One of the problems of using limestone as a porous sorbent for SO_2 is that since the molar volume of $CaSO_4$ (52.2 cm^3/mol) is greater than that of CaO (16.9 cm^3/mol) and since no particle expansion occurs during sulfation the maximum possible extent of limestone conversion is about 50 % (Hartman and Coughlin 1976; Oka 2004). Moreover since material at the pore entrances sulphates before that in the internal porous structure the entrances become blocked with $CaSO_4$ resulting in a maximum conversion of somewhat less than 50 %. Hence to achieve high levels of sulphur retention it is necessary to operate with quantities of sorbent well in excess of the stoichiometric amount. This excess is normally expressed in terms of the ratio of moles of calcium in the stone to moles of sulphur in the fuel. Highley (1975) derived the empirical relationship:

$$R = 100[1 - \exp(-M\beta)] \tag{3.8}$$

for the percentage reduction in SO_2 emission, R, in terms of the Ca/S mole ratio, β, and a constant M whose value depends on coal type, additive and operating conditions. The influence of Ca/S mole ratio for a commercial CFBC is shown in Fig. 3.13.

Under atmospheric pressure conditions in a FBC the optimum temperature for sulfation is in the range 800–850 °C the decreasing rate at higher temperatures

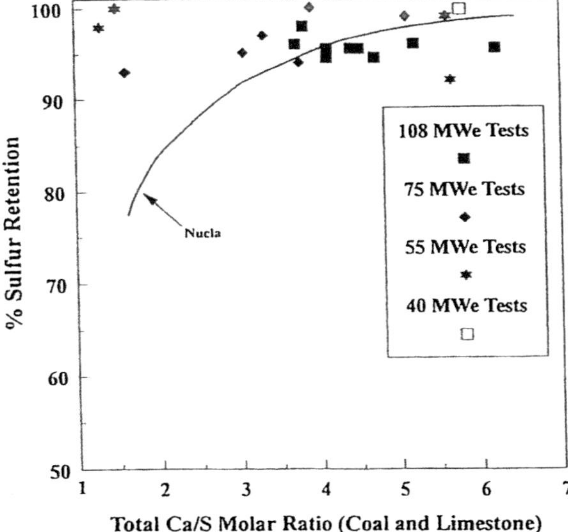

Fig. 3.13 Influence of the Ca/S ratio on sulphur capture (Brereton 1997)

being considered to be due to the regeneration of SO_2 by reaction with CO (Lyngfelt and Leckner 1989; Hansen et al. 1991):

$$CaSO_4 + CO \rightarrow CaO + SO_2 + CO_2$$

At elevated pressures sulfation rates with limestone decrease due to the displacement of the equilibrium of the calcination reaction to the left:

$$CaCO_3 \leftrightarrow CaO + CO_2$$

Sulfation with dolomite however is relatively unaffected by increased pressure (Borgwardt and Harvey 1972; Roberts et al. 1975). A sulphur-capture efficiency of 90 % has been achieved in large-scale CFBC's with Ca/S mole ratios of 1.8–3.1 but owing to the problem of pore plugging up to 50 % of the individual sorbent particle remains unsulfated. The utilization of sorbents under operating conditions depends on a number of interacting factors such as limestone type and its attrition characteristics, fluid-bed hydrodynamics and the efficiency of the installed cyclones and a number of options for increasing the utilization have been investigated including:

- Physical grinding of particles
- Rehydration of spent sorbents with steam or water
- Slurrying and reinjection of ash
- The ADVACAT process involving the production of reactive hydrated calcium silicates for cold-end absorption of SO_2 (Basu 2006).

In the second process the objective is to rehydrate the spent sorbent to calcium hydroxide which when reinjected into the furnace decomposes to expose fresh sorbent surface. An experimental study of the process has been reported by Couturier et al. (1994).

3.7.4 Sulfation Models

In non-catalytic gas-solid reactions such as the sulfation of limestone as reaction proceeds the solid phase is composed of a reacted shell separated from the unreacted core by a moving boundary. Simple models of these processes such as the "shrinking core" model assume the overall conversion to be a function of three elements:

- gas-phase mass transfer of reactant to the solid-particle surface
- diffusion through the porous reacted shell
- chemical reaction at the boundary between the reacted and unreacted zones.

In many cases the inter-zone boundary is not sharply defined but diffuse and the reaction rate depends on features of the solid structure such as its porosity. To model these processes more realistically Szekely and Evans (1970, 1971) introduced the "grain" model in which the reacting pellet is assumed to consist of uniformly spherical particles or grains of radius r_s separated by pores through which the reactant gas diffuses. If the centres of adjacent spheres are spaced a distance l apart the porosity of the pellet, S_g, is given by:

$$S_g = 1 - \frac{4\pi}{3}\frac{r_s^3}{l^3}N \tag{3.8}$$

where each space element of volume l^3 contains N spherical particles of radius r_s. The factor N allows for denser packing with a range of particle sizes. The assumptions on which the original model was based are:

- external gas-phase mass transfer is neglected
- chemical reaction is of the first order
- the system is isothermal
- the initial structure is maintained throughout (this assumption was subsequently modified—see below).

For spherical grains located in the ith row from the external surface the rate of change of core radius is:

$$-\frac{dr_i}{dt} = \frac{C_{pi}}{\rho\left[\frac{r_i}{D_s} - \frac{r_i^2}{D_s r_s} + \frac{1}{k}\right]};0 \le r_i \le r_s \tag{3.9}$$

r_i = radius of unreacted grain core in row i
r_s = initial grain radius
D_s = diffusivity of the reactant gas in the solid product
C_{pi} = reactant gas concentration in the pores
ρ = molar density of the solid
k = first-order rate coefficient

The initial conditions for Eq. (3.9) are:

$$r_i = r_s \, at \, t = 0 \tag{3.10}$$

The diffusivity of the reactant gas in the product solid, D_s, was defined in terms of the gas-phase diffusivity of the reactant, D_p, the porosity of the solid, S_g and a tortuosity factor F_T given a value of 2.75:

$$D_s = \frac{S_g D_p}{F_T} \tag{3.11}$$

The authors conceded that the most serious limitation of the grain model was the neglect of structural changes as reaction proceeds. This question was addressed by Georgiakis et al. (1979) in the following way.

For the general reaction scheme between a gas A and a solid B giving a solid product P and a gas product Q:

$$\mathrm{A(g) + bB(s) \rightarrow pP(s) + gQ(g)}$$

the rate of change of grain size is:

$$\frac{dg_2}{dt} = -\frac{\beta C}{\beta + g_2\left(1 - \frac{g_2}{g_1}\right)} \tag{3.12}$$

$$g_1^3 = \alpha + (1-\alpha)g_2^3 \tag{3.13}$$

where g_1 and g_2 are the dimensionless radii of the grains and their unreacted cores respectively (Fig. 3.14):

$$g_1 = \frac{r_1}{r_0};\ g_2 = \frac{r_2}{r_0}$$
$$\beta = \frac{D_s}{kr_0}$$
$$C = \frac{c_A}{c_{A0}}$$
$$\alpha = \frac{pv_P}{bv_B}$$

Here c_A and c_{AO} are the concentrations of reactant gas in the pores and the bulk phase respectively; v_P and v_B are the molar volumes of product and reactant solids respectively. For values of α less than unity the porosity increases with time; for values greater than unity, as in the case of limestone sulfation, the porosity decreases with time. The local conversion at time t and position R is:

$$x(R,t) = 1 - g_2^3(R,t) \tag{3.14}$$

from which the overall conversion is:

$$X(t) = 3\int_0^1 R^2 x(R,t)dR \tag{3.15}$$

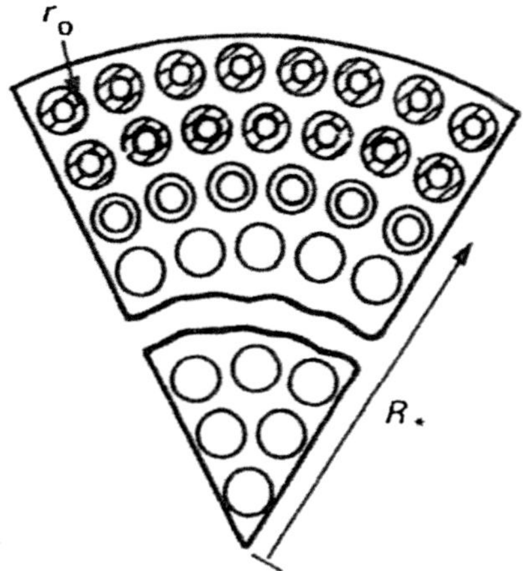

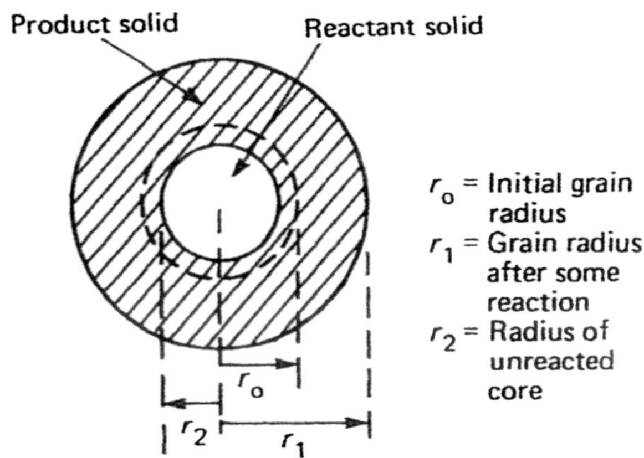

Fig. 3.14 Schematic representation of changing-grain-size model (Georgiakis et al. 1979)

Neglecting external mass transfer Eq. (3.12) can be integrated to:

$$\int_{1}^{g_2} \left[\beta + g_2\left(1 - \frac{g_2}{g_1}\right)\right] dg_2 = -\beta t \tag{3.16}$$

which leads to:

$$t = (1 - g_2) + \frac{1}{2\beta}\left(1 - g_2^2\right) - \frac{1}{2\beta(\alpha - 1)}\left(g_1^2 - 1\right) \tag{3.17}$$

The overall conversion, $X(t)$, may be found from Eqs. (3.12) to (3.17) and expressed as a function of the local porosity, ε, where:

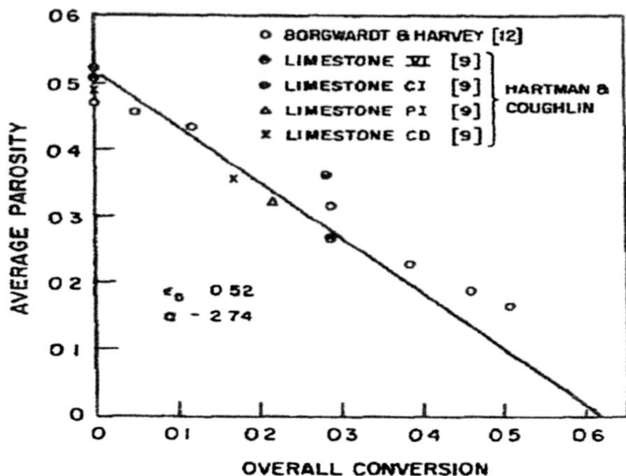

Fig. 3.15 Dependence on the average porosity of overall conversion (Georgiakis et al. 1979)

$$g_1^3 = \frac{(1-\varepsilon)}{(1-\varepsilon_0)} \tag{3.18}$$

ε_0 being the initial porosity. Comparison with experimental values determined by Borgwardt and Harvey (1972) and Hartman and Coughlin (1974) (Fig. 3.15) shows excellent agreement.

Basu (2006) developed a simplified model for sulphur capture in terms of the ratio of feed rates of sorbent and fuel, the weight fraction of calcium carbonate in the stone, the weight fraction of sulfur in the coal and the cyclone collection efficiency. The model was developed with a view to demonstrating the effects on desulfurization of feedstock characteristics and various design and operating parameters.

3.7.5 *Nitrogen Oxides*

The oxides of nitrogen, NO, NO_2 and N_2O, originate predominantly from the nitrogen compounds contained in the fuel, the oxidation of atmospheric nitrogen in these systems being limited by their low temperature of operation. The two oxides NO and NO_2, normally lumped together as NO_x, are major pollutants from coal-fired power stations and their emissions are strictly controlled by legislation in many countries. Interest in the formation of N_2O in combustors has grown in recent years owing to its effect as a greenhouse gas more powerful in absorbing IR radiation than carbon dioxide (Elkins 1989).

The nitrogen content of most coals is of the order of 1–2 % on a dry mineral-free basis and is more or less equally distributed between the volatile component and the char. The oxidation reactions leading to the formation of NO_x and N_2O are numerous and complex involving both direct oxidation and secondary conversion via ammonia and HCN (Basu 2006). Some 77 % of the fuel nitrogen is oxidised to NO_x, a proportion of the remainder being reduced to N_2 by char and CaO (Johnsson 1989); N_2O is mainly formed by the oxidation of HCN (Moritomi and Suzuki 1992).

A number of methods are available for reducing NO_x emissions. One such is air staging where a proportion of the combustion air is introduced downstream leaving the NO_x to be reduced by char in the lower region. This and other methods such as ammonia injection have been discussed by Basu (2006).

3.8 Gasification

Processes for converting coal into gas have a history stretching back to the early nineteenth century. The product of these early processes was town gas, a blend of the volatile constituents of coal with water gas formed by blowing steam at atmospheric pressure through hot beds of the devolatilized char leaving coke as a valuable by-product. The plants were generally built on a large scale and produced a gas containing a wide range of impurities such as hydrogen sulphide, organic sulphur compounds and unsaturated hydrocarbons. Town gas had a calorific value of 20–24 MJ/m^3 and as a result of being about 50 % hydrogen was characterised by a high flame speed. The main gasification reactions are:

$$\begin{aligned} C + H_2O \rightarrow CO + H_2; &\quad \Delta H = 131.3\,\text{kJ/mol} \\ C + CO_2 \rightarrow 2CO; &\quad \Delta H = 172.5\,\text{kJ/mol} \\ CO + H_2O \rightarrow CO_2 + H_2 &\quad \Delta H = -41.2\,\text{kJ/mol} \end{aligned}$$

In the 1920s and 30s a number of new processes were developed with the objective of gasifying all the organic matter in coal and leaving ash as the only solid residue. The coal particles, prepared in pretreatment steps, were kept in motion during reaction by being fluidized, as in the Winkler process, dust injected, as in the Koppers-Totzec process or suspended in a molten slag as in the Rummel process. The most successful of these processes was that developed by the Lurgi company in Germany. Here the gasifier operates under pressure of 2–3 MPa and uses a steam/oxygen mixture as the gasifying medium. The feedstock is lump coal fed to the reactor via a pressurized hopper and kept in motion during reaction by means of a rotating grate through which the ash is discharged. Lurgi gasifiers continue to be used on a large scale e.g. in the South-African based Synthol process (Sect. 2.2.6) where some 84 reactors produce a total of 55 million Nm^3/day of syngas (Higman and van der Burgt 2008).

In recent decades interest has shifted towards integrated gasification combined cycle (IGCC) plants where, after cleaning, the generated syngas is fed to a combustion turbine for electricity generation (Fig. 3.16).

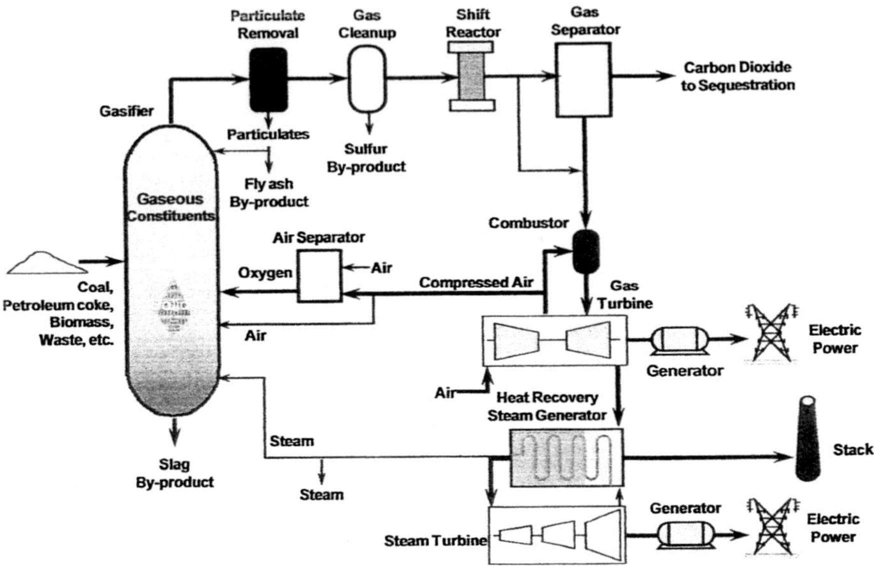

Fig. 3.16 IGCC system with carbon capture (Breault 2010)

Here the gasifier configurations are generally of three types: moving bed, entrained bed and fluidized bed, some characteristics of the various processes being listed in Table 3.3.

Fluidized-bed gasifiers operate at temperatures below the softening point of ash, typically 950–1100 °C for coal and 800–950 °C for biomass and have historically operated with peat or biomass or with low-rank coals such as lignite. However the greater carbon efficiency demonstrated by entrained-flow designs which operate at high temperatures (1300–1700 °C) and pressures (15–25 bar) make them preferable to fluidized beds for advanced IGCC power cycles (Dry and Beeby 1997).

Table 3.3 Gasifier configurations and processes (Adapted from Higman and van der Burgt 2008)

	Moving bed		Fluidized bed		Entrained flow	Transport flow
Ash	Dry	Slagging	Dry	Agglomerate	Slagging	Dry
Coal feed	6–50 mm		6–10 mm		<100 μm	6–10 mm
Coal rank	Any	High	Low	Any	Any	Any
Gas Temp./°C	425–650		900–1050		1250–1600	900–1000
Process	Lurgi	BGL	Winkler, HTW, KRW		Shell, GEE, E-Gas	KBR
			U-Gas, HRL		KT, Siemens, MHL	

The fluidized-bed Winkler process was patented in 1922 and began operation at the BASF plant at Leuna, Germany, in 1926. In the reactor section of the original process dried coal, sized 0–10 mm, was screw fed from a storage bunker into the gasifier where it was fluidized with air (and/or oxygen) and steam at atmospheric pressure and a temperature, depending on the type of coal used, in the range 850–1000 °C. During reaction the larger agglomerated ash particles settle to the bottom of the bed and are removed by a screw discharger; the lighter ash particles are elutriated with the product gas and unreacted coal fines. The unreacted coal particles are gasified by a secondary steam/air mixture injected into the freeboard but the efficiency is not high and some 20 % of the carbon fed can be lost with the fly ash. Some 70 Winkler reactors were built and operated up to the mid-1980s but most have now been shut down on economic grounds (Higman and van der Burgt 2008). An improved version of the process, the so-called High Temperature Winkler or HTW process, was developed by Rheinbraun AG in the 1980s. Operating at an elevated pressure of up to 30 bar the bed is maintained at 800 °C to avoid ash fusion and elutriated particles are separated via a cyclone and returned to the main reactor. The HTW process produces a better quality gas than the original Winkler and is suitable for lignite and biomass (Basu 2006).

3.8.1 Gasification Modelling

A number of models of fluidized-bed gasification of solid fuels have been published. Hamel and Krumm (2001) developed a model for gasification in bubbling beds and used it to simulate four units of different scales from atmospheric laboratory scale to pressurised commercial scale processing brown coal, peat and sawdust. Good agreement was found with published experimental data.

Ross et al. (2005) described an isothermal and a non-isothermal model used to simulate the performance of a full-scale fluidized-bed coal gasifier, a 2MWth pilot plant and a laboratory-scale bed. The temperature profile of the gas phase predicted by the non-isothermal model followed the trend of the of bubble temperature due to a majority of the product gas flowing through the bed as bubbles. The temperature of the cold feed gas was predicted to rise in the lower region of the bed and to reach a peak in higher regions due to homogeneous combustion. Better agreement with experimental data was found with the non-isothermal model than with the isothermal variant.

Brown et al. (2010) as part of a study of chemical looping combustion (see below) developed a bubbling two-phase model of a gasifier based on the following assumptions:

- the bed was isothermal throughout
- the bubble phase was free of solid particles
- gasification of the lignite char by CO_2 and reduction of the hemeatite (Fe_2O_3) oxygen carrier occurred solely in the emulsion phase which was maintained at the point of incipient fluidization
- the emulsion phase was well mixed while the bubble phase was in plug flow the flow being enhanced by the gas expansion due to gasification of the solid.

The model predicted the experimentally observed reaction between char and Fe_2O_3 at 800 °C and 1 bar pressure which was shown to be entirely due to the gasification products, the solid-solid reaction occurring at a negligible rate if at all.

3.9 Chemical Looping

Concern about the steadily increasing concentration of carbon dioxide in the Earth's atmosphere and its resulting effect on global warming has prompted the development of a number of novel processes aimed at reducing or removing CO_2 from anthropogenic sources. Chemical-looping combustion (CLC) and gasification are examples of these in which a carbon-containing fuel such as coal, syngas or a hydrocarbon is burned not with air but with a solid oxygen carrier such as a metal oxide, Me_xO_y, to produce a nitrogen-free effluent gas containing a high concentration of CO_2 in a form, after condensing out the water component, suitable for subsequent capture and storage:

$$(2n + m)M_xO_y + C_nH_{2m} \rightarrow (2n + m)Me_xO_{y-1} + mH_2O + nCO_2 \quad \text{(a)}$$

The depleted oxygen carrier is then regenerated with air in a second stage:

$$Me_xO_{y-1} + {}^1\!/\!_2\, O_2 \rightarrow Me_xO_y \quad \text{(b)}$$

The two processes are conveniently carried out in two coupled fluidized-bed reactors the fuel reactor and the air reactor between which the oxygen carrier is circulated. Depending on the nature of the oxygen carrier reaction (a) is frequently endothermic; reaction (b) is always exothermic.

The history of CLC per se dates from the 1980s but the principle had been established earlier by Lewis and Gilliland in a process designed to produce pure carbon dioxide (Lewis and Gilliland 1954). Ishida introduced the term "Chemical-Looping Combustion" for the capture of CO_2 using iron- and nickel-based oxygen carriers (Ishida et al. 1987) and in the subsequent period several groups around the world have taken the technique forward. The historical development of the field has been comprehensively reviewed by Adanez et al. (2012) and by Fan (2010).

The single most important element in any CLC system is the solid oxygen carrier. It should have a combination of favourable chemical and physical characteristics amongst which are the following:

- good activity in both reduction and oxidation reactions
- high oxygen-carrying capacity
- suitable thermodynamics vis-a-vis the fuel to be combusted
- stability under repeated redox cycles
- good mechanical strength and fluidizability
- high melting point and resistance to agglomeration
- low cost

Over 700 different materials have been examined in this respect (Fan and Li 2010; Lyngfelt et al. 2008; Li and Fan 2008), transition metals and their oxides being found to be the most promising particularly the combinations Ni/NiO, Cu/CuO, Fe/FeO, Fe_3O_4/Fe_2O_3 and MnO/Mn_3O_4. Mattisson et al. (2006) examined the thermodynamic aspects of nickel oxide as an oxygen carrier for the oxidation of methane. For the fuel reactor:

$$4NiO + CH_4 \rightarrow 4Ni + 2H_2O + CO_2; \quad \Delta H = 134.4\,kJ/mol$$

and for the air reactor:

$$4Ni + 2O_2 \rightarrow 4NiO; \quad \Delta H = -239.7\,kJ/mol$$

The results showed that Ni/NiO can convert over 99 % of methane to CO_2 at temperatures below 900 °C decreasing to 97.7 % at 1200 °C, the balance being made up of CO and H_2. Mattisson and Lyngfelt (2001) found similar results for the copper, iron and manganese combinations shown above (Fig. 3.17).

Despite the favourable thermodynamics, owing to its tendency to agglomerate at high temperatures, unsupported NiO performs poorly over repeated redox cycles (Jin and Ishida 2004) and must be supported on other materials to perform adequately; alumina, Al_2O_3, has been much studied in this context (e.g. Sedor et al. 2008) its main functions being to provide a high dispersion of the metal and increase the fluidization properties and mechanical strength of the NiO. The agglomeration tendencies, relatively low melting points and other disadvantages of the pure materials make it imperative to operate all CLC systems with supported oxygen carriers and a number of other supports such as SiO_2,TiO_2, ZrO_2 and yttria-stabilized-zirconia (YSZ) have been investigated – good summaries of this work are given by Hossain and de Lasa (2008) and Adanez et al. (2012).

The redox reactions indicated above are somewhat idealized and in practice the CLC process can be influenced by side reactions and by the presence of impurities in the fuel gas. Carbon formation via pyrolysis:

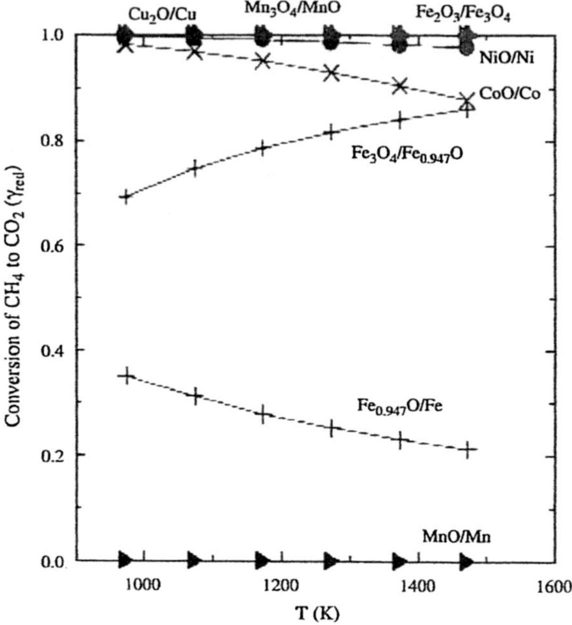

Fig. 3.17 Conversion of CH_4 to CO_2 as a function of temperature for different oxygen carriers (Mattisson and Lyngfelt 2001)

$$CH_4 \rightarrow C + 2H_2; \quad \Delta H = 88\,\text{kJ/mol}$$

and the reverse Boudouard reaction:

$$2CO \rightarrow C + CO_2; \quad \Delta H = -72.5\,\text{kJ/mol}$$

can lead to the deposition of coke on the metal-oxide surface. Cho et al. (2005) reported a study of carbon formation on NiO and Fe_2O_3 in a laboratory-scale fluidized-bed reactor and found that for the nickel-based carrier significant carbon formation only occurred when more than 80 % of the available oxygen had been consumed i.e. carbon formation correlated with low fuel conversion. No carbon was formed on the iron oxide-based carrier. The authors conclude that in practice carbon formation should not be a problem since the process should be run under conditions of high fuel conversion.

The presence of sulphur-containing compounds in potential fuel gases such as refinery gas and syngas can lead to deactivation of the oxygen carrier through the formation of sulphides or sulphates on the surface. The influence of sulphur compounds was investigated by Jerndal et al. (2006) who concluded that to maintain carrier efficiency fuel gases may need to be desulfurized prior to the combustion stage.

The kinetics of the reduction and oxidation reactions have been studied in terms of the "shrinking-core" model referred to above under *Sulfation models*. In the present case as the reduction reaction proceeds the metal-metal oxide interface

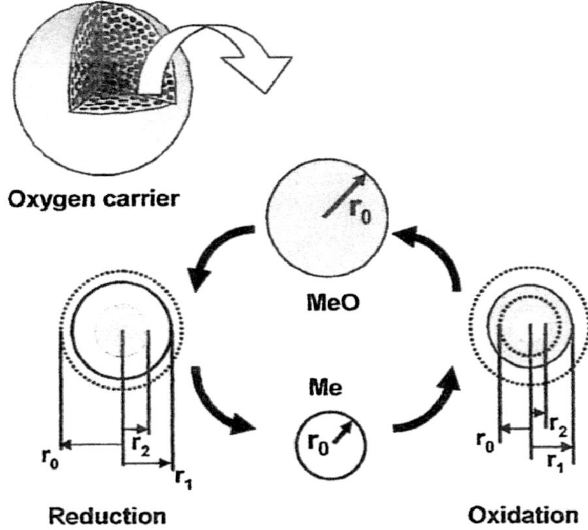

Fig. 3.18 Scheme of the changing-grain-size model (Garcia-Labiano et al. 2005)

moves through the grains towards the centre of the pellet leaving behind a porous layer through which the fuel gas diffuses Fig. 3.18).

Unlike in the case of limestone desulfurization however the molar volume of the product metal is less than that of its oxide so the porosity of the product layer increases with time and as there are no external mass-transfer limitations the kinetics are controlled by the intrinsic reaction step. The opposite applies for the oxidation stage where the rate of reaction is controlled by diffusion of gas through the product layer (Ryu et al. 2001). Garcia-Labiano et al. (2005) studied the reactions of different oxygen carriers based on Cu, Co, Fe, Mn and Ni during reduction with three fuel gases CH_4, CO and H_2 and subsequent oxidation with O_2. Here the oxidation reactions were always exothermic causing an increase in particle temperature whereas the reduction reactions could be exothermic or endothermic depending on the metal oxide and fuel gas. A non-isothermal changing-grain-size model was applied and predictions made of the profiles of oxygen-carrier conversion and temperature inside the carrier particles. The study concluded that under typical CLC conditions with particle sizes lower than 0.3 mm and 40 wt% metal oxide content the particles could be considered to be iso-thermal with little likelihood of particle sintering.

A number of sub-pilot scale CLC units using gaseous fuels have been reported in the literature: 10 kWth units at Chalmers University of Technology, Sweden (Lyngfelt and Thunman 2005) and the Institute of Carboquimica, Spain (Adanez et al. 2006), a 50 kWth unit at KIER, Korea (Ryu et al. 2010) and a 140 kWth unit at the Vienna University of Technology (Kolbitsch et al. 2009). All these units are alike in being made up of two interconnected fluidized beds but differ in their details of design and operation. In the Chalmers University system (Fig. 3.19) the

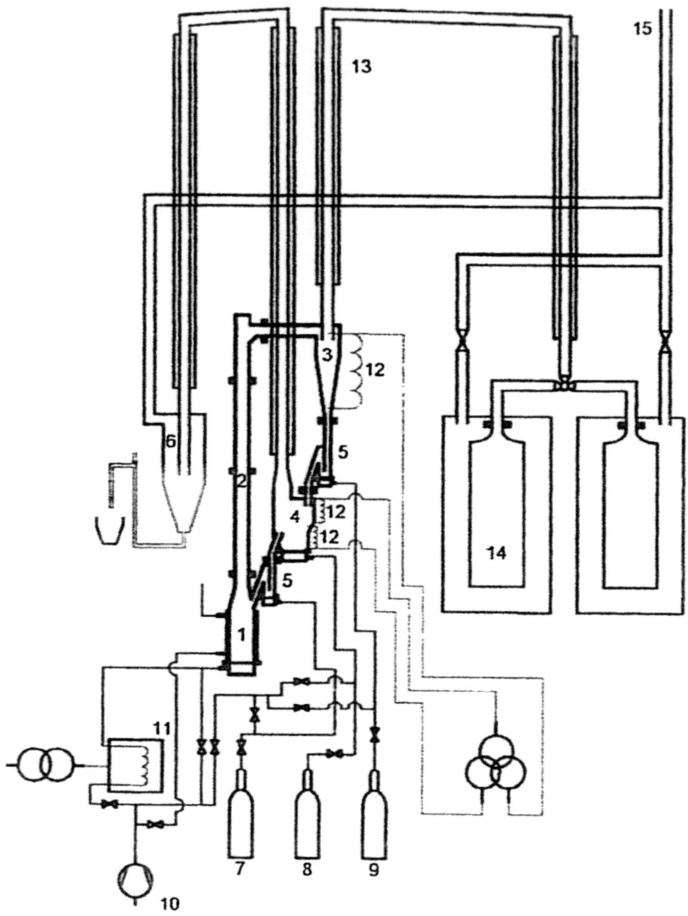

Fig. 3.19 Chalmers University 10 kW CLC unit (Linderholm et al. 2008). *1* Air reactor, *2* Riser, *3* Cyclone, *4* Fuel reactor, *5* Upper and lower particle locks, *6* Water seal, *7* Nitrogen, *8* Natural gas, *9* Nitrogen, *10* Air, *11* Preheater, *12* Heating coils, *13* Finned tubes, *14* Filters, *15* Connection to chimney

oxygen-depleted solids from the bubbling-bed fuel reactor are fed by gravity to the air-blown reactor and riser via a loop seal that prevents gas leakage between the reactors.

Entrained solids leaving the top of the riser are separated in a cyclone before being fed back to the fuel reactor via a second loop seal. The diameters of the fuel and air reactors are 0.25 and 0.14 m respectively and that of the riser 0.072 m; the heights of the three components are 0.35, 0.53 and 1.86 m respectively. In the reported test runs with methane as the fuel gas velocities in the fuel reactor were in the range 5–15 times the U_{mf} of the oxygen carrier, a material made up of NiO supported on $NiAl_2O_4$ with a mean particle diameter of 100–200 μm; in these runs

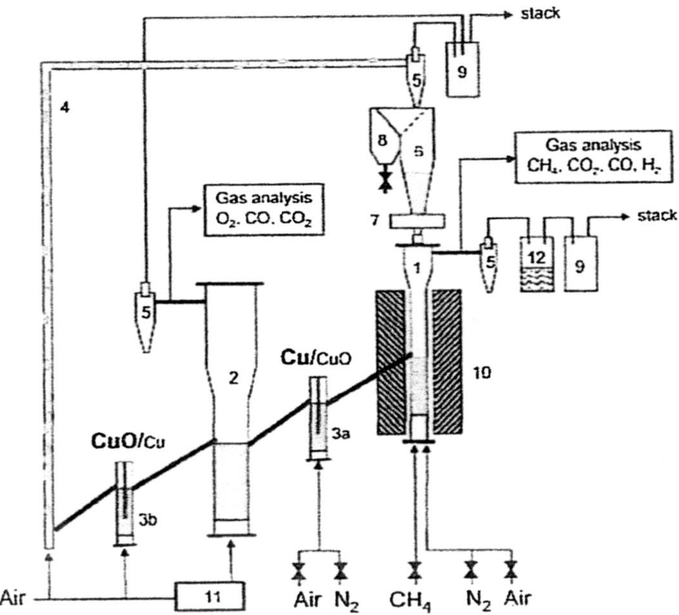

Fig. 3.20 10 kW CLC unit at CSIC-ECB in Zaragoza (Adanez et al. 2006). 1. Fuel reactor, 2. Air reactor, 3. Loop seals, 4. Riser, 5. Cyclone, 6. Solids reservoir, 7. Solids valve, 8. Diverting solids valve, 9. Filters, 10. Oven, 11. Air preheater, 12. Water condenser

a 99 % conversion to CO_2 was achieved at fuel-reactor temperatures in the range 660–950 °C (Linderholm et al. 2008).

The unit built at the Instituto de Carboquimica is shown schematically in Fig. 3.20. This also consists of two interconnected fluidized beds separated by loop seals. The fuel reactor is a bubbling bed 0.108 m in diameter and 0.5 m high with an extended freeboard of 1.5 m; the air reactor is 0.16 m in diameter and 0.5 m high with an extended height of 1.0 m. The oxygen carrier was 14 % CuO on Al_2O_3 prepared by dry impregnation, two particle sizes of 100–300 μm and 200–500 μm being used with circulation rates of 60–250 kg/h at bed temperatures of 800 °C. This temperature was lower than that often used with other materials due to the tendency of copper-based carriers to agglomerate at higher values (Cho et al. 2004). The unit was operated for a 200 h period with methane as fuel, complete conversion to CO_2 being obtained with oxygen carrier-to-fuel ratios, φ, in excess of 1.4 where:

$$\varphi = \frac{F_{CuO}}{4F_{CH_4}} \tag{3.19}$$

Here F_{CuO} and F_{CH_4} are the molar flow rates of oxygen carrier and fuel respectively. Particle attrition rates were high at the start of the run but decreased to a constant 0.04 wt%/h after 40 h of operation (Adanez et al. 2006) .

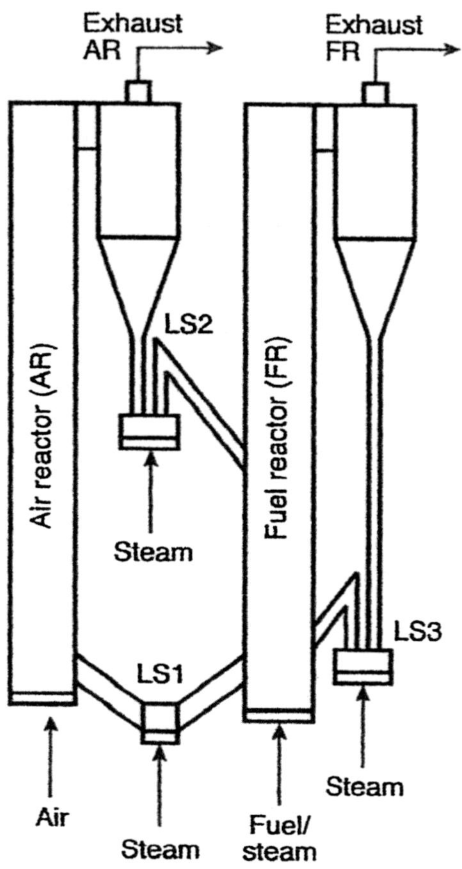

Fig. 3.21 140 kW unit at Vienna University of Technology (Lyngfelt 2011)

The 140 kW unit at the Vienna University of Technology is shown in outline in Fig. 3.21 (Pröll et al. 2009). It consists of a fuel reactor 0.158 m in diameter and 3 m high, an air reactor 0.15 m in diameter and 4.1 m high, upper and lower loop seals, an internal loop seal and two cyclones. The fuel reactor operates at 850 °C in turbulent flow and the air reactor at 940 °C in the dilute or dense-phase transport regime. Fuels tested have been hydrogen, methane, syngas and propane with nickel-based oxygen carriers and ilmenite ($FeTiO_3$) a potentially low-cost material. With a NiO-based oxygen carrier of mean particle diameter 135 μm and a 65 kg solids inventory both CH_4 conversion and CO_2 yield were in excess of 90 %.

The studies referred to above were all concerned with gaseous fuels but the widespread availability and use of solid fuels such as coal makes the application of CLC technology to solids of considerable importance. A major problem in this case however is that of separating particles of unreacted char and reduced oxygen carrier before the latter are transferred to the air reactor for regeneration. Any char entering the air reactor would be oxidized to CO_2 and leave mixed with nitrogen thereby defeating the object of the whole process. A possible way round the problem would be to gasify the solid fuel to syngas (CO + H_2) in a separate reactor before admitting the product gas to the CLC system. However the production of a suitable syngas free from nitrogen requires the use of pure oxygen which would necessitate the incorporation of an air-separation unit again thereby negating the benefits of the technique. An alternative approach is to carry out the gasification with steam and CO_2 in the fuel reactor itself alongside the action of the fluidized oxygen carrier. This approach has been studied by a number of groups (e.g. Scott et al. 2006; Leion et al. 2007; Berguerand and Lyngfelt 2008; Shen et al. 2009; Fan 2010; Thon et al. 2013) with varying degrees of success.

The technology of chemical looping combustion and gasification is currently at an early stage of development and much work needs to be done before the technique is ready for large-scale industrial implementation. A number of problems remain unresolved. In the reported gaseous-fuel systems despite the high fuel conversions and CO_2 yields achieved a significant amount of fuel gas remains unconverted in the fuel reactor exit gases necessitating the incorporation of a "polishing" step where the unconverted fuel is burned with oxygen. In the solid-fuel systems conversions are generally significantly lower than in the case of gaseous fuels and the problem of fuel carryover to the air reactor remains an issue. Reaction between oxides of sulphur and nitrogen with the oxygen carrier and its interaction with the residual ash from the combustion preclude the use of more expensive materials making the use of cheaper but less active carriers such as ilmenite an important consideration. It may be concluded as Lyngfelt has written that "large efforts to scale up more established CO_2 capture technologies may divert interest from a technology which is quite new, less well known and which may need a longer time period to reach commercial size" (Lyngfelt 2011).

References

Adanez J, Gayan P, Garcia-Labiano F, de Diego LF (1994) Axial voidage profiles in fast fluidized beds. Powder Tech 85(1):259–268

Adanez J, Gayan P, Celaya J, de Diego LF, Garcia-Labiano F, Abad A (2006) Chemical looping combustion in a 10 kWth prototype using CuO/Al_2O_3 oxygen carrier: effect of operating conditions on methane combustion. Ind Eng Chem Res 45:6075–6080

Adanez J, Abad A, Garcia-Labiano F, Gayan P, de Diego LF (2012) Progress in chemical looping combustion and reforming technologies. Prog Energy Comb Sci 38:215–282

Agarwal PK (1986) A single particle model for the evolution and combustion of coal volatiles. Fuel 65:803–810

Anderson KD, Manaker AM, Jr Stephans EA (1997) Operating experience of the Tennessee Valley Authority's 160 MW atmospheric fluidized combustion unit. In: Proceedings of fourteenth international conference on fluidized bed combustion. ASME, New York, pp 39–45

Avedesian MM, Davidson JF (1973) Combustion of carbon particles in a fluidized bed. Trans I Chem E 51:121

Basu P (2006) Combustion and gasification in fluidized beds. Taylor & Frances, Florida

Basu P, Halder PK (1989) Combustion of single carbon particles in a fast fluidized bed of fine particles. Fuel 68:1056–1063

Basu P, Broughton J, Elliott DE (1975) Coal combustion in fluidized beds. Inst Fuel Symp Ser No 1 1(A3)

Berguerand N, Lyngfelt A (2008) Design and operation of a 10 kWth chemical looping combustor for solid fuels—Testing with South African coal. Fuel 87:2713–2726

Bishop JW (1970) Proceedings second international conference on fluidized bed combustion paper. US-EPA, pp 4–4

Borghi G, Sarofim AF, Beer JM (1977) A mechanistic model of coal combustion in fluidized beds. AIChE 70th annual meeting, New York

Borgwardt RH, Harvey RD (1972) Properties of carbonate rock related to sulphur dioxide activity. Environ Sci Tech 6:350

Breault RW (2010) Gasification processes old and new: a basic review of the major technologies. Energies 3:216–240

Brereton C (1997) Combustion performance. Chapter 10 in circulating fluidized beds. In: Grace JR, Avidan A, Knowlton TM (eds) Blackie academic & professional, London

Brown TA, Dennis JS, Scott SA, Davidson JF, Hayhurst AN (2010) Gasification and chemical-looping combustion of a lignite char in a fluidized bed of iron oxide. Energy Fuels 24:3034–3048

Carlson GA, Mitchell RF (1971) Method of chlorinating titanium bearing minerals. U S Patent 3591333

Castleman III JM (1985) Process performance of the TVA 20 MW atmospheric fluidized-bed plant. In: Proceedings of eighth international conference on fluidized bed combustion. ASME, New York, pp 196–207

Cessaroli B, Lohne O (2011) Solar grade silicon feedstock. In: Luque A, Hegedus S (eds) Handbook of photovoltaic science and engineering. Wiley, New York (Chapter 5)

Cho P, Mattisson T, Lyngfelt A (2004) Comparison of iron-, nickel-, copper- and manganese-based oxygen carriers for chemical-looping combustion. Fuel 83:1225–1245

Cho P, Mattisson T, Lyngfelt A (2005) Carbon formation on nickel- and iron oxide-containing oxygen carriers for chemical looping combustion. Ind Eng Chem Res 44:668–676

Coutourier MF, Marquis DL, Steward FR, Volmeranger Y (1994) Reactivation of partially sulfated limestone from a CFB combustor by hydration. Canadian J Chem Eng 72:91–97

Davidson JF (2000) Circulating fluidized bed hydrodynamics. Powder Tech 113:249–260

Dennis JS, Hayhurst AN (1990) Mechanism of the sulfation of calcined particles in combustion gases. Chem Eng Sci 45:1175–1187

Dry RJ, Beeby CJ (1997) Applications of CFB technology to gas-solid reactions. In: Grace JR, Avidan AA, Knowlton TM (eds) Circulating fluidized beds. Blackie A&M, London (Chapter 12)

Elkins JW (1989) State of the research for atmospheric nitrous oxide (N_2O). In: Paper contributed to the International Panel on Climate Change, Boulder, Colorado

Elkins TS (1997) Process for controlling the temperature of a fluidized-bed reactor in the manufacture of titanium tetrachloride. U S Patent 5670121

Elliot DE (1970) Proceedings of second international conference on fluidized bed combustion: Paper 0-1. US-E{PA

Fan LS (2010) Chemical-looping systems for fossil-energy conversions. Wiley-AIChE, New York

Fan LS, Li F (2010) Chemical looping technology and its fossil energy conversion applications. Ind Eng Chem Res 49:10200–10211

Garcia-Labiano F, de Diego LF, Adanez J, Abad A, Gayan P (2005) Temperature variations in the oxygen carrier particles during their reduction and oxidation in a chemical-looping combustion system. Chem Eng Sci 60:852–862

Georgiakis C, Chang CW, Szelely J (1979) A changing grain size model for gas-solid reactions. Chem Eng Sci 34:1072–1075

Glaeser HH, Spoon MJ (1995) Fluidized-bed process for chlorinating titanium-containing materials and coke useful in such processes. U S Patent 5389353

Glicksman LR (1997) Heat transfer in circulating fluidized beds. In: Grace JR, Avidan AA, Knowlton TM (eds) Circulating fluidized beds. Blackie Academic and Professional, London (Chapter 8)

Grace JR (1990) High-velocity fluidized bed reactors. Chem Eng Sci 45(8):1953–1966

Greenwood NN, Earnshaw A (1997) Chemistry of the elements, 2nd edn. Butterworth Heinemann, Oxford

Hamel S, Krumm W (2001) Mathematical modelling and simulation of bubbling fluidized-bed gasifiers. Powder Tech 120:105–112

Hansen PFB, Dam-Johansen K, Bank LH, Ostergard K (1991) Sulfur retention on limestone under fluidized-bed combustion conditions—an experimental study. In: Anthony EJ (ed) Proceedings of 11th international conference on fluid bed combustion, pp 281–291. ASME New York

Hartge E-W, Li Y, Werther J (1986). Analysis of the local structure of the two-phase flow in a fast fluidized bed. In: Basu P (ed) Circulating fluidized bed technology. Pergamon, Toronto, pp 153–160

Hartman M, Coughlin RW (1974) Reaction of sulphur dioxide with limestone and the influence of pore structure. Ind Eng Chem Proc Des Dev 13:248–253

Hartman M, Coughlin RW (1976) Reaction of sulphur dioxide with limestone and the grain model. AIChE J 22:490–498

Highley J (1975) A model of coal combustion in a fluidized bed. Inst Fuel Symp Ser No. 1, Vol 2: 37

Higman C, van der Burgt M (2008) Gasification, 2nd edn. Gulf Publications, London

Horio M (1997) Hydrodynamics. In: Grace JR, Avidan AA, Knowlton TM (eds) Circulating fluidized beds. Blackie Academic and Professional, London (Chapter 2)

Horio M, Morishita K (1988) Flow regimes in high velocity fluidization. Jpn J Multiphase Flow 2:117–136

Hossain MM, de Lasa HI (2008) Chemical-looping combustion for inherent CO_2 separations—a review

Ishida M, Zheng D, Akehata T (1987) Evaluation of a chemical-looping-combustion-power-generation system by graphic-exergy analysis. Energy 12:147–154

Jazayeri B (2003) Applications for chemical production and processing. In: Yang W-C (ed) Handbook of fluidization and fluid-particle systems. Marcel Dekker, New York (Chapter 16)

Jerndal E, Mattisson T, Lyngfelt A (2006) Thermal analysis of chemical-looping combustion. Chem Eng Res Des 84:795–806

Jin H, Ishida M (2004) A new type of coal gas fuelled chemical-looping combustion. Fuel 83:2411–2417

Johnsson JE (1989) A kinetic model for NO_x formation in fluidized-bed combustion. In: Manaker A (ed) Proceedings of 10th international conference on fluid bed combustion. ASME, New York, p 1112

Johnsson F (2007) Fluidized bed combustion for clean energy. In: Bi X, Berruti F, Pugsley T (eds) Engineering Conferences International, Fluidization XII, New York, pp 47–62

Johnsson F, Leckner B (1995) Vertical distribution of solids in a CFB furnace. In: Proceedings of 13th ASME international conference on fluidized bed combustion, pp 61–79

Johnsson F, Andersson S, Leckner B (1991) Expansion of a freely bubbling fluidized bed. Powder Tech 68(2):117–123

Kolbitsch P, Pröll T, Bolhar-Nordenkampf J, Hofbauer H (2009) Design of a chemical-looping combustor using a dual circulating fluidized-bed reactor system. Chem Eng Technol 32:398–403

Kunii D, Levenspiel O (1991) Fluidization engineering, 2nd edn. Butterworth-Heinemann, Boston

La Nauze RD (1985) Fundamentals of coal combustion. In: Davidson JF, Clift R, Harrison D (eds) Fluidization. Academic Press, London (Chapter 19)

Leckner B (1998) Fluidized-bed combustion: mixing and pollutant limitation. Prog Energy Comb Sci 24:31–61

Leckner B, Szentanni P, Winter F (2011) Scale-up of fluidized-bed combustion—a review. Fuel 90:2951–2964

Lee YY (1997) Design considerations for CFB boilers. In: Grace JR, Avidan A, Knowlton TJ (eds) Circulating fluidized beds. Blackie Academic and Professional, London (Chapter 11)

Leion H, Mattisson T, Lyngfelt A (2007) The use of petroleum coke as fuel in chemical-looping combustion. Fuel 86:1947–1958

Lewis WK, Gilliland ER (1954) Production of pure carbon dioxide. US Patent 2665972

Li F, Fan L-S (2008) Clean coal conversion processes—progress and challenges. Energy Environ Sci 1:248–267

Linderholm C, Abad A, Mattisson T, Lyngfelt A (2008) 160 h of chemical-looping combustion in a 10 kW reactor system with a NiO-based oxygen carrier. Int J of Greenhouse Gas Control 2:520–530

Louge M (1997) Experimental techniques. In: Grace JR, Avidan AA, Knowlton TM (eds) Circulating fluidized beds. Blackie Academic and Professional, London (Chapter 9)

Luckos A, den Hoed P (2004) The carbochlorination of titaniferous feedstocks in a fluidized bed. In: Engineering Conferences International, Fluidization XI, New York, pp 555–562

Lyngfelt A (2011) Oxygen carriers for chemical-looping combustion—400 h of operational experience. Oil Gas Sci Technol 66:161–172

Lyngfelt A, Leckner B (1989) Sulfur capture in fluidized-bed combustors: temperature dependence and lime conversion. J Inst Energy 62(450):62–72

Lyngfelt A, Thunman H (2005) Construction and 100 hr of operational experience of a 10 kW chemical-looping combustor. In: Thomas DC, Benson SM (eds) Carbon dioxide capyure for storage in deep geologic formations. Elsevier, Oxford

Lyngfelt A, Johansson M, Mattisson T (2008) Chemical-looping combustion—status of development. In: 9th International Conference on Circulating Fluidized Beds, Hamburg

Mattisson T, Lyngfelt A (2001) Capture of CO_2 using chemical-looping combustion. In: Proceedings of 1st Biennial Meeting of the Scandanavian-Nordic Section of the Combustion Institute, April 18–20, Göteborg, Sweden

Mattisson T, Johansson M, Lyngfelt A (2006) The use of NiO as an oxygen carrier in chemical-looping combustion. Fuel 85:736–747

Moritomi H, Suzuki Y (1992). Nitrous oxide formation under fluidized-bed cobbustion conditions. In: Potter O, Nicklin DJ (eds) Proceedings of 7th international conf on fluidization, United Eng Foundation, New York, pp 495–507

Muschelknautz U, Muschelknautz E (1991) Special design of inserts and short entrance ducts to recirculating cyclones. In: Proceedings of international conference on circulating fluidized beds, pp 597–602

Newby RA (2003) Applications for gasifiers and combustors. In: Yang W-C (ed) Handbook of fluidization and fluid-particle systems. Marcel Dekker, New York (Chapter 15)

Oka SN (2004) Fluidized bed combustion. Marcel Dekker, Basel

Palchonok GI, Breitholtz C, Thunman H, Leckner B (1997) Impact of heat and mass transfer on combustion of a fuel particle in CFB boilers. In: Proceedings of 14th international conference on fluidized bed combustion, pp 871–888

Pallares D, Johnsson F (2006) Macroscopic modelling of fluid dynamics in large-scale circulating fluidized beds. Prog Energy and Comb Sci 32:539–569

Park D, Levenspiel O, Fitzgerald TJ (1980) A comparison of the plume model with currently used models for atmospheric fluidized bed combustors. Chem Eng Sci 35:295–301

Pröll T, Kolbitsch P, Bolhar-Nordenkampf J, hofbauer H (2009) A novel dual circulating fluidized-bed system for chemical-looping systems. AIChE J 55:3255–3266

Rhodes MJ, Geldart D (1987) A model for the circulating fluidized bed. Powder Tech 53(3):155–162

Rhodes MJ, Zhou S, Hirama T (1991) Effects of operating conditions on longitudinal solids mixing in a circulating fluidized bed riser. AIChE J 37:1450–1458

Roberts AG, Stanton JE, Wilkins DM, Beacham B, Hog HR (1975) Combustion in fluidized beds. Inst fuel Symp Ser No. 1 1:D4

Ross IB, Davidson JF (1982) The combustion of carbon particles in a fluidized bed. Trans IChemE 60:108

Ross DP, Yan H-M, Zhong Z, Zhang D-K (2005) A non-isothermal model of a bubbling fluidized-bed coal gasifier. Fuel 84:1469–1481

Ryu H-J, Bae D-H, Han K-H, Lee S-Y, Jin G-T, Choi J-H (2001) Oxidation and reduction characteristics of oxygen-carrier particles and reaction kinetics by unreacted-core model. Korean J Chem Eng 18:831–837

Ryu H-J, Jo S-H, Park Y, Bae D-H, Kim S (2010) Long-term operational experience in a 50 kWth chemical-looping combustor using natural gas and syngas as fuels. In: Proceedings of 1st International conference on Chemical Looping, Lyon, France

Sabino MEL, Passos ML, Branco JR (2013) Integrated fluidized-bed reactors for silicon production. In: Passos ML, Barrozo MAS, Mujumdar AS (eds) Fluidization engineering: practice (Chapter 6)

Scott SA, Dennis JS, Hayhurst AN, Brown T (2006) In situ gasification of a solid fuel and CO_2 separation using chemical looping. AIChE J 52:3325–3328

Sedor KE, Hossain MM, de Lasa HI (2008) Reactivity and stability of Ni/Al_2O_3 oxygen carrier for chemical-looping combustion. Chem Eng Sci 63:2994–3007

Shen L, Wu J, Xiao J (2009) Experiments on chemical-looping combustion of coal with a NiO-based oxygen carrier. Combust Flame 156:721–728

Sinclair JL (1997) Hydrodynamic modelling. In: Grace JR, Avidan AA, Knowlton TK (eds) Circulating fluidized beds. Blackie Academic and Professional, London (Chapter 5)

Skinner DG (1970) The fluidized combustion of coal. Mills and Boon, London

Szekely J, Evans JW (1970) A structural model for gas-solid reactions with a moving boundary. Chem Eng Sci 25:1091–1107

Szekely J, Evans JW (1971) A structural model for gas-solid reactions with a moving boundary—II. The effect of grain size, porosity and temperature on the reaction of porous pellets. Chem Eng Sci 26:1901–1913

Takahashi M, Nakabayashi Y, Kimura N (1995) The 350 MWe Takehara plant. VGB Kraftwerkstech 75:427–434

Thon A, Kramp M, Hartge E-U, Heinrich S, Werther J (2013) Operation of a coupled fluidized-bed system for CLC of solid fuels with a synthetic Cu-based oxygen carrier. In: Kuipers JAM, Mudde RF, van Ommen JR, Deen NG (eds) Fluidization XIV, pp 63–70. Eng Conf Int, New York

Turnbaugh DT, Morris AJ, Perkins JB (2007) Method for the analysis of gas produced by a titanium tetrachloride fluidized-bed reactor. U S Patent 7183114

Werther J (2005) Fluid dynamics, temperature and concentration fields in large-scale CFB combustors: 1–25. In: Kofa C (ed) Circulating fluidized bed technology, vol VIII. International Academic Publishers, Beijing, pp 1–25

Werther J, Hirschberg B (1997) Solids motion and mixing. In: Grace JR, Avidan MM, Knowlton TM (eds) Circulating fluidized beds. Blackie A&P, London (Chapter 4)

Winkler J (2003) Titanium dioxide. Vincentz Network, Hanover

Yates JG (1983) Fundamentals of fluidized-bed chemical processes. Butterworths, London

Zhang XY (1980) Proceedings of sixth international conference on fluidized bed combustion. Conf 800428, US Dept of Energy, Morgantown, pp 36–40

Chapter 4
Conversion of Biomass and Waste Fuels in Fluidized-Bed Reactors

Abstract Thermal conversion processes, namely combustion, gasification and pyrolysis are presented in this chapter. The conversion mechanisms that take place in fluidized-bed reactors are explained with emphasis on materials in-feeding, devolatilization and volatile conversion, char conversion and particle attrition and elutriation. A comparison between conventional and unconventional fuels is made, with particular focus on gasification of waste fuels for energy generation. The operating parameters in gasification processes are discussed, with reference to feeding (feeding methods, number of feeding points and solid fuel feed size), bed depth, bed temperature, fluidization velocity, equivalent ratio, and inerts content. To conclude, examples of current technologies which employ fluid-bed combustors, pyrolysers and gasifiers in operation in the UK and around the world are discussed. Advanced thermal processes such as fluid-bed plasma for waste gasification are also touched upon.

4.1 Introduction

The thermal conversion of new generation fuels in fluidized-bed reactors is discussed in this Chapter. Generally, such types of materials are classified as low grade fuels because their calorific value is relatively low compared to fossil fuels, or because they are in some way difficult to treat in conventional equipment. In some instances, for example, they can give rise to emission problems if treated conventionally, Doug Orr (2000). Such materials may occur naturally (e.g. biomass) or originate as waste materials (municipal solid waste, commercial and industrial waste, automotive shredder residues, etc.).

Given the rising concerns surrounding energy security and the increasing political emphasis on environmental sustainability, there are growing support schemes promoting the capabilities of waste-to-energy power generation in fluidized beds, Defra (2013). Waste-to-energy, or energy-from-waste, is the process of generating energy in the form of electricity and/or heat from thermal treatment of low grade fuels.

J.G. Yates and P. Lettieri, *Fluidized-Bed Reactors: Processes and Operating Conditions*, Particle Technology Series 26,
DOI 10.1007/978-3-319-39593-7_4

Table 4.1 lists a number of the more common types of such fuels along with indications of their calorific value, ash content and water content. This list is intended merely to be illustrative and is in no way complete as the variety of waste fuels is almost endless.

The thermal treatment of low grade and waste fuels may be carried out for a variety of reasons; they include:

- the generation of heat, steam or power;
- the production of syngas (CO, H_2, CH_4) as precursors for chemical synthesis and advanced fuels generation.
- the treatment of a waste material to lessen its toxic properties;
- the utilisation of a low grade energy source to reduce fossil fuel demands or because there is no easier alternative;

changing of the physical form of a waste material (into pulverised ash for example) to make it more readily or economically disposable; it may even be possible to convert the waste material to a usable product (e.g. road paving material, construction material, etc.); low grade and waste fuels are often very variable in quality. This characteristic can make them difficult to use in conventional

Table 4.1 Classification of some types of low grade and waste fuels (BCURA 2005)

Fuel type	Calorific value (as received) MJ/kg	Ash (as received) % w/w	Moisture (as received) % w/w
Municipal solid waste	9–21	15–30	20–30
combustible components:			
– paper, cardboard	12–18	6	5
– wood	16–21	5–9	0.5
– textiles	15–20	0.5–3	2–10
– sawdust – plastics	18 17–40	1 10	3 2
– vegetables, garbage	18	16	70
– rubber, leather	18–26	10	1–10
Refuse derived fuel (RDF)	10	50	33
Industrial wastes			
– oil	19–30		
– paints	19–30		
– food processing	17–36		
– plastics	17–40		
– rubber, tyres	20–30		
Agricultural waste			
– wood chips	13	3.3	30
– paper	23	14	4.6
– sawdust	17	6	7

equipment but normally presents no problems for fluidized-bed conversion systems provided they can be fed to the reactor at a controllable rate and without interruptions in supply caused by variations in their physical nature. These fuels also vary enormously in their physical properties, calorific value and chemical composition as can be seen from Table 4.1. It is difficult, therefore, to give specific, quantitative design information for individual fuels. A generalised process scheme is then outlined and the factors to be considered at design stage are discussed. Finally, examples of advanced thermal technologies for specific application in the UK and in the world are given to illustrate the use of fluidized-bed reactors.

4.2 Thermal Conversion Approaches

Thermal conversion processes have been known and used for centuries. Despite this long history, development of advanced thermal conversion technologies for processing wastes has only become a focus of attention in recent years stimulated by the search for more efficient energy recovery and better ways of disposal.

Combustion, also known as mass burn incineration, is defined as the total thermal conversion of a substance with sufficient oxygen to oxidise the fuel completely. The general characteristic of combustion of a waste stream is that an excess air is required to ensure complete oxidation (maximum temperatures typically above 1000 °C); the fuel is completely oxidised to carbon dioxide and steam, leaving only a small amount of carbon in the ash (less than 3 % by weight of ash). Furthermore, all of the chemical energy in the fuel is converted into thermal energy, leaving no unconverted chemical energy in the flue gas and very little unconverted chemical energy in the ash.

Gasification is the partial oxidation of a substance with lower quantities (i.e. sub-stoichiometric) of oxygen compared to full combustion. Temperatures are typically above 750 °C, and products are gas (main combustible components being methane, hydrogen, and carbon monoxide) and a solid residue (consisting of non-combustible material and a small amount of carbon). The overall process does not convert all of the chemical energy in the fuel into thermal energy but instead leaves a big portion of the chemical energy in the syngas and, to a minor extent, in the solid residues (char). The typical net calorific value of the gas from gasification using oxygen is 10–15 MJ/Nm3.

Pyrolysis is the thermal degradation of a substance in the absence of added oxygen, at relatively low temperatures (typically from 300–600 °C). Products are oils and/or char (i.e. solid residue of non-combustible material and a significant amount of carbon), with synthetic gas as by-product. The latter consists in a mixture of carbon monoxide, hydrogen, methane and some longer chain hydrocarbons including tars. The char residue from pyrolysis processes could contain up to 40 % carbon representing a significant proportion of the energy from the input waste. Recovery of the energy from the char is therefore important for energy efficiency.

A wide variety of biomass and waste fuels can be handled successfully by fluidized-bed reactors and this process is considered to be more versatile than other more conventional forms of technologies, Scala (2013). Fuels in any physical form, and which may vary considerably in quality, can be converted successfully in a fluidized-bed reactor through the three approaches described above. The main characteristics of fluidized-bed reactors which make this high degree of versatility possible are as follows:

1. Rapid rates of mixing of the bed solids coupled with a relatively high thermal capacity of the bed solids. As a result considerable fluctuations in the fuel properties will not upset the process.
2. A relatively low reaction temperature which means that problems of volatilisation or of fusion of waste products are minimised. There is also a possibility that certain toxic heavy metal elements may be retained by the bed solids which can make their subsequent safe disposal easier, Williamson et al. (1994).
3. The ease of removal of the bed solids from a fluidized-bed reactor compared with conventional stoker or grate firing of solid fuels means that fuels with high inerts content are readily handled.
4. A fluidized-bed reactor can be operated, depending on the feed material, either autothermally or with auxiliary fuel injection. The auxiliary fuel may be natural gas or any other available material of reasonable calorific value (e.g. tyres and plastics).
5. The process can be operated to reduce emissions of tars to technically acceptable low values by the addition of solid sorbents.
6. The variety of feeding methods available makes it possible to feed fuel in any physical form and, if so desired, fuels in more than one physical form may be fed simultaneously.

4.3 Conversion Mechanisms in Fluidized-Bed Reactors

4.3.1 Fluidization and the Reacting Environment

In fluidized beds, the solid fuel and the bed material are intimately mixed and suspended on upward-blowing bubbles of fluidising gas. The upwards and sideways coalescing movement of bubbles provides intense agitation and mixing of the bed particles, which make fluid beds ideal for applications where high mass and heat transfer rates are required. In such systems (i.e. combustors or gasifiers), the particles are initially heated to above the ignition temperature of the fuel to be burned and then combustion/gasification takes place when the fuel is delivered into or onto the heated fluidized particles. The fuel burns (completely or partly) by virtue of the oxygen within the fluidising gas (air, oxygen, steam-oxygen or enriched-oxygen air), which is delivered by a fan (or blower) through the distributor plate and upwards through the bed particles. During steady-state gasification, temperature is controlled by the opposing effects of the heat input from the burning fuel in the bed, versus outgoing heat in the devolatilised gases and further heat 'consumed' by endothermic reactions in the gas-phase (e.g. steam reforming, water gas-steam carbon, etc.).

When considering gasification or combustion in fluidized beds it is important to remember that the oxidants leaving the distributor flows through the bed in two ways. According to the two-phase theory (Grace and Clift 1974), one portion—smaller—produce the dense phase, flows through the bed at the minimum fluidising velocity, and has good contact with the bed solids. The other portion, which in fluidized-bed gasifiers has a volume flow varying from 1.5 to 3 times that of the dense phase flow, forms the bubble phase. When they are first formed at the base of the bed the bubbles have good contacting with the dense phase but they rapidly coalesce as they rise up through the dense phase. The larger bubbles mix to only a limited extent with the dense phase and provide a gas flow which to some extent short circuits the bed and has a much shorter residence time than the dense phase.

Various refractory materials can be used to form the original 'bed' of particles, the most convenient being graded sand, around 1 mm in mean diameter, enabling fluidising velocities in the range 1–3 m/s. Alternatively, graded limestone or dolomite can be used if sulphur capture and/or tar reforming are required.

The produced gas leaving the freeboard is then forced to flow through a cyclone to separate part of the elutriated material (e.g. fly ash, unreacted char, bed particles, etc.), and cooled to (say) 200 °C before going through the gas cleaning.

Making a technical survey of fluidized-bed thermal conversion processes involves the analysis of several other phenomena, which include:

- Materials in-feeding
- Devolatilization and volatile conversion
- Char conversion
- Particle attrition and elutriation

A more detailed description of the process in a fluid bed reactor follows in this Section.

4.3.2 *Material In-Feeding*

Fuel feeding in fluidized beds is one of the basic problems that has to be solved in order to achieve efficient conversion. Poor mixing of inert bed particles and fuel in horizontal direction, and short fuel particle residence time are two of the main drawbacks reported for bubbling fluidization at large scale, Gomez-Barea and Leckner (2010). In this sense, the way in which waste fuel is injected to the fluid bed reactor is a very critical point.

In-bed fuel feeding has been the first and most employed feeding system for fluidized-bed coal boilers in the past. Fuel is crushed to a size of ~5 mm, and dried to ensure moisture content not higher than 6–8 %, prior to transport and feeding, Oka (2004). Material is fed into the vicinity of the distributor plate from hoppers

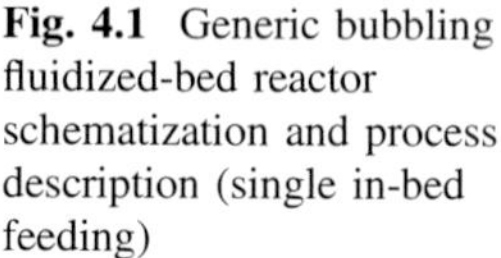

Fig. 4.1 Generic bubbling fluidized-bed reactor schematization and process description (single in-bed feeding)

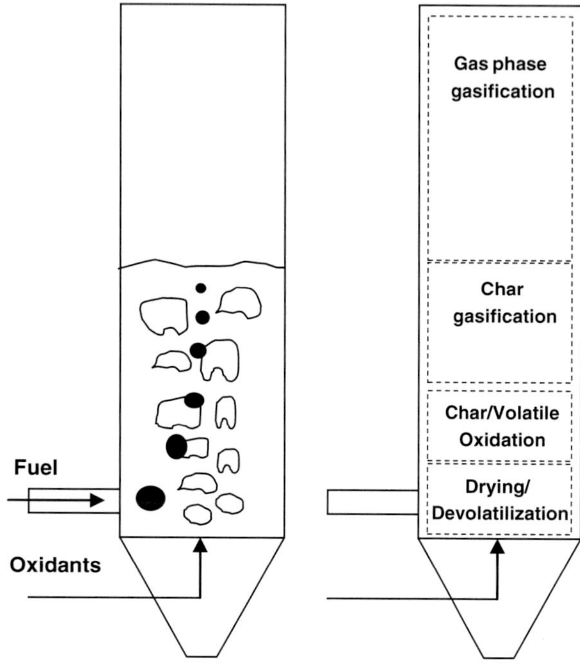

(Fig. 4.1), connected to screw augers via a rotary valve or a pneumatic conveying system, with slight nitrogen over-pressure to stop back feeding of producer gas and bed material through the in-feed. The basic advantage of this type of fuel feeding is the large residence time of fuel particles in the bed and higher conversion efficiency, Oka (2004). However, with operation on waste, the following issues are of concern:

- If the gasifier runs at greater than atmospheric pressure, there is potential for migration or leakage of hot gas into the fuel feed mechanism.
- Waste material is heterogeneous, both in composition and morphology. Bridging in hoppers and blocking of screw feeds are common issues with wastes. It is reported by Vreugdenhil (2010) that bridging in the feeding line was the main cause that resulted in abandoning pure waste feeding in circulated fluidized-bed gasifiers. Blockages may also be caused by large inerts in waste fuels. Drying and pelletisation are common techniques to provide a more homogeneous fuel form to aid feeding, and also by definition remove any large inerts above the pellet size. However this comes at high costs and energy penalties.
- The high levels of volatiles, especially plastics in municipal solid waste are likely to lead to rapid devolatilisation at the base of the gasifier, on contact with the hot bed material. This could exacerbate the issue of gas transport back up the in-feed and could lead to melting of the feed in the auger and exacerbate risk of blockages

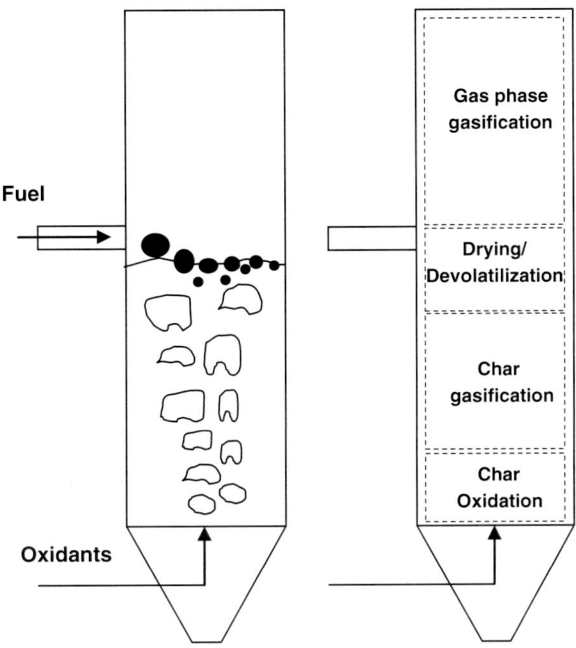

Fig. 4.2 Generic bubbling fluidized-bed reactor schematization and process description. Single over-bed feeding (*top*)

A different option is to inject the fuel directly over-bed by means of screw feeders (Fig. 4.2). These feeding devices were also developed for conventional fluid bed combustor boilers, and have been successfully applied in fluid bed gasification. A general advantage of over-bed feeding systems is the possibility of feeding moist and gross fuels "as received", with a particle range between 0–50 mm, although it is recommended that percentage of particles smaller than 0.5 mm should not be higher than 10 % for this type of systems, Oka (2004). The deficiencies of this type of feeding are its limits in terms of the need to specify a particular particle size range and the likelihood of elutriation of small particles and their subsequent conversion above the bed (rather than inside the bed). A further general deficiency of this type of system is the requirement for recycling/disposal of unreacted fuel particles, given the high elutriation losses that are likely to be experienced (see Sect. 4.3.5).

4.3.3 *Devolatilization and Volatile Conversion*

When the temperature exceeds 250 °C, the fuel particle organic matter starts to thermally degrade, with the detachment of the volatile matter from the solid fuel matrix (being 'char'), Basu and Kaushal (2009). This step is usually referred to as pyrolysis (or devolatilization), wherein water vapour, organic liquids and non-condensable gases, such as CO, H2, CO2, are separated from the solid carbon and ash content of the fuel. The vapour/liquid product comprises mostly of hydrocarbons and tar (i.e. dark, oily, viscous material, consisting mainly of heavy organic and mixed oxygenates). In absence of excess oxygen, the volatiles and char undergo a second conversion step and they modify their composition due to the occurrence of several reactions becoming a syngas (see Table 4.2).

Most of these reactions are endothermic and require a consistent amount of energy to proceed. This is usually supplied by the exothermic reactions within the same environment. Since the initial devolatilization is a rapid process, it has a negligible effect on the overall conversion time. Nevertheless, the location of devolatilization significantly affects the heat release profiles throughout the reactor.

On entering the hot fluidized bed most solid fuels initially decompose rapidly into a volatile portion which enters the gas phase and a solid portion or "char" which remains with the inert bed solids (Fig. 4.1). In view of the distribution of the air in a fluidized bed described above it can be seen that efficient in-bed conversion of gaseous products, and also of liquid and solid fuels producing a high proportion of volatiles, will only be obtained by feeding them at a relatively large number of points low down in the bed. Otherwise the volatiles will tend to react preferentially in the freeboard. This will occur, either if the volatiles enter the bubble phase, as the bubbles will by-pass the dense phase of the bed, or if the volatiles enter the dense phase, as the latter will tend to become oxygen deficient. In gasification, a modest proportion of freeboard combustion is acceptable although it is more efficient to burn the char within the bed rather than other volatiles in the freeboard, in which CO and H_2 are the most reactive. Excessive freeboard reaction, however, may result in an inability to maintain the desired bed temperatures if the heat release in the bed becomes less than the sum of the heat removed from it as sensible heat and as heat required for endothermic reactions, BCURA (2005).

Table 4.2 Typical reaction scheme, Materazzi et al. (2013)

Reaction name	Biomass gasification	Energy (kJ/mol)
Exothermic:		
Combustion	(Char/Volatiles) $C + O_2 \rightarrow CO_2$	−398.3
Partial oxidation	(Char/Volatiles) $C + 1/2\,O_2 \rightarrow CO$	−123.1
Water gas shift	$CO + H_2O \leftrightarrow H_2 + CO_2$	−40.9
CO methanation (I)	$CO + 3H_2 \leftrightarrow CH_4 + H_2O$	−217.0
CO methanation (II)	$2CO + 2H_2 \leftrightarrow CH_4 + CO_2$	−257.0
Endothermic:		
Pyrolysis	Biomass $\rightarrow$ Char + Volatiles + $CH_4 + CO + H_2 + N_2$	+ 200–400
Methane steam reforming	$CH_4 + H_2O \leftrightarrow CO + 3H_2$	206.0
Water gas/steam carbon	(Char/Volatiles) $C + H_2O \rightarrow CO_2 + H_2$	118.4
Boudouard	(Char/Volatiles) $C + CO_2 \rightarrow 2CO$	159.9

4.3.4 "Char" Conversion and Fuel Reactivity

As the solid carbon-containing particles formed on initial decomposition of the fuel (or fed if the fuel is non-volatile), mix with the inert bed solids, heterogeneous reactions continue until the particle, is either burnt completely, or is discharged with the bed material for bed height control, or is removed from the bed by elutriation, BCURA (2005). This situation is likely to be the same for all fuels that have initially, or can form, carbonaceous 'char' particles irrespective of whether they are fossil or low grade or waste fuels in origin. The rate of reaction at the surface of a char particle will depend on the local air supply (determined by fluidizing conditions and the number of feed points for a fixed total feed), on the chemical activity of the carbon forming the char, and on the bed temperature, Basu and Kaushal (2009) and Scala (2013).

The chemical activity of the char varies according to the kind of fuel and how the char is formed. For example, it is well known that the chars derived from hard wood biomass are less reactive than those derived from soft wood biomass, Kersten et al. (2005).

4.3.5 Particle Attrition and Elutriation

This already complex picture is further complicated in a fluidized-bed reactor by the parallel ash release and comminution phenomena that can remarkably change the structure of mother fuel particles and then strongly affect its conversion process, Chirone et al. (1991) and Gomes-Barea et al. (2010).

If the fuel has only a very low, friable, ash content, the fuel ash is mainly degraded by the action of the fluid bed, such that it is substantially carried away i.e. elutriated (as 'fly ash'), within the emergent flue gases. Alternatively, if a high ash material is treated, especially one that leaves behind coarse particles of metals, glass or adventitious stone, some of the ash (named 'bottom ash') remains in the bed. If such ash is of similar size to the original bed particles, the ash will fluidise and eventually coexist with the original bed particles.

The process through which particle size decreases and ash is released includes the following four steps: primary and secondary fragmentation, attrition by abrasion and percolative fragmentation (Fig. 4.3). Primary fragmentation (Chirone et al. 1991) occurs immediately after the injection of the fuel particle into the bed, as a consequence of thermal stress caused by rapid heating and by volatile release. It generates coarse particles whose size and shape are influenced by fuel properties such as volatile content and swelling index. Secondary fragmentation and attrition by abrasion (Chirone et al. 1991) are determined by mechanical stress due to collisions between particles and with the furnace interior: the former generates coarser and non-elutriable fragments while the latter generates finer and elutriable

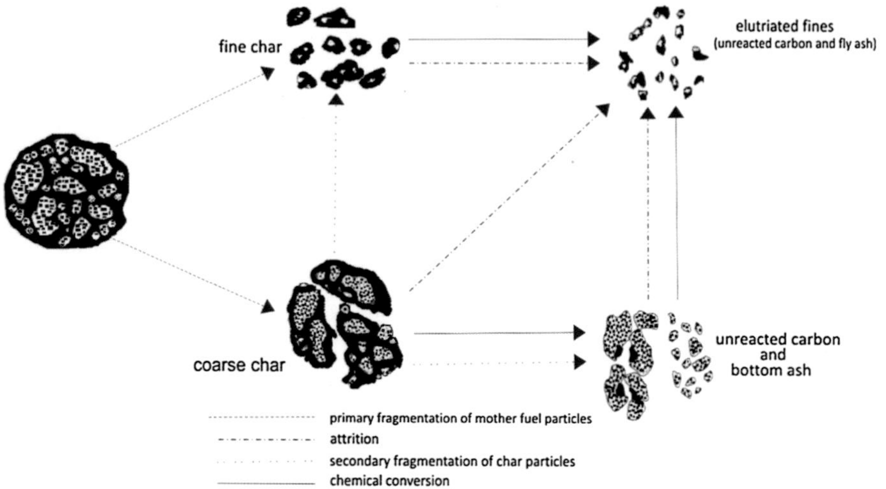

Fig. 4.3 Schematic of the series-parallel comminution phenomena of a solid fuel particle during a gasification process in a fluidized-bed reactor, Scala (2013)

fragments. Secondary fragmentation is favoured by swelling index, particle size and oxygen concentration increases. The oxidation conditions in the bed control attrition by abrasion, too. Experimental measurements reveal that, after an initial period of high attrition, the generation of fine particles reaches a stationary value proportional to the excess gas velocity above the minimum for fluidization and to the total carbon surface exposed in the bed, Ross and Davidson (1981).

The phenomenon that most affects carbon conversion efficiency is elutriation. This is the process in which fine particles are carried out of a fluidized bed due to the fluid flow rate passing through the bed. Fine particles that are subjected to elutriation can directly come from the fuel fed to the fluid bed gasifier or, alternatively, they can be produced during the gasification process and attrition.

4.3.6 Comparison Between Conventional and Waste Fuels

Gasification of waste fuels in fluidized beds includes a wide variety of non-fossil solid materials, ranging from mixed plastic to municipal, agricultural and industrial waste. Although there is a certain amount of operating experience in connection with this topic, (Saxena and Jotshi 1994 and Anthony 1995), a complete comprehension of the phenomena occurring during fluidized-bed treatment of these fuels, refuse-derived fuels in particular, is still lacking, probably due to the great difference in physical and chemical features from conventional ones. In fact, waste fuels are characterized by high moisture and volatile content, a porous and fragile structure, a low density and high intrinsic reactivity. The potential of a large quantity of moisture in fuel particles amplifies drying time and postpones devolatilization. On the contrary, low moisture and the high volatile contents lead to shorter devolatilization times and larger quantities of volatiles evolved: as a result, a larger contribution to the overall heat release is associated with drying process and homogeneous volatile reactions. Refuse-derived fuels devolatilization is completed at (or close to) the bed surface and a large fraction of the volatile matter is released directly in the freeboard: a direct consequence of bypassing the bed is that the post-conversion of volatiles in the splashing region leads to significant local overheating with respect to the bed, Scala (2013). Besides, fine carbon particles are significantly formed by attrition and fragmentation of coarse particles: this feature reflects the propensity of such fuels to give rise to either friable chars or even to a multitude of fragments of very small size. As a result, the conversion of fixed carbon occurs as much through the generation of fines, followed by their conversion over their residence time in the bed, as through direct conversion of coarse char particles. Because of high reactivity, the fine char particles are mostly burned in the bed, Anthony (1995). Whereas conventional fuels like coal undergo moderate primary fragmentation: after devolatilization about 99 % of the fixed carbon can be found in coarse char particles. Consequently, coal conversion occurs primarily in the bed, mostly via coarse char particle direct combustion.

The high quantity of fly ash and volatile material in wastes can also provide a decrease in thermal output, create high ash clinkering, and increase emission of tars and particulates, Materazzi et al. (2015). In fact, these reactors need to be operated at lower temperatures to prevent sintering of the ashes causing defluidization of the bed and, consequently, tend to produce a syngas containing high levels of condensable organics and gaseous hydrocarbon species which can be problematic in subsequent stages. Furthermore, the large quantities of gases and vapours leaving the solid matrix can entrain organic and inorganic material, even if the material itself is non-volatile, thus producing a large amount of residues downstream. As a result, the combination of high velocities in fluidized-bed reactors and high volatile matter in waste fuels indicates a potential for creating significant tar condensation and fly ash deposition problems during thermal treatment, with the severity varying significantly with the different nature of the feedstock, Miles at el. (1996).

4.4 Operating Parameters

4.4.1 Feeding Methods

The choice of the feeding method may often be a key factor in the design of a fluidized-bed gasifier to handle a low grade or waste fuel. It will often be desirable to carry out experimental tests to determine the optimum feeding method, BCURA (2005).

When choosing a feeding method the following general points should be borne in mind.

- Light and/or small sized particles which are readily elutriated are best fed near the base of the bed to give the maximum opportunity for reacting within the bed. It may be advantageous to feed such materials premixed with a suitable carrier.
- Solids with a high volatiles content (e.g. municipal solid waste, mixed plastics, etc.) are best fed near the base of the bed at a relatively large number of inlet points so that good initial mixing between air and fuel is achieved.
- Over-bed feeding methods will give more in-bed gasification if the fuel is deposited just above the bed surface rather than dropped from a height.
- Loosely adherent material like sludges may tend to flow erratically with lumps falling into the bed. Such lumps are often easily broken up by the particle movement in the bed and may not present a problem in maintaining a state of good fluidization.

4.4.2 Equivalence Ratio

The oxygen used in a process determines the products and temperature of the reaction. The oxygen consumed is typically plotted as the equivalence ratio, i.e. the oxygen used relative to that required for complete combustion. A very low or zero oxygen use is indicative of pyrolysis, shown at the left of the figure. An equivalence ratio (SR) of about 0.25 is typical of the gasification region at the middle; and combustion is indicated by a SR = 1 (Fig. 4.4).

A further consequence of the division of the air into two portions, as described above, is that it is impossible to obtain complete combustion of any fuel in a fluidized bed using a stoichiometric supply of air. It is recommended that, in general, the equivalence ratio level should not be higher than 15 % for any fuel (i.e. SR = 1.15). For some biomass fuels, however, acceptable combustion can be obtained at an excess air level of 10 % (SR = 1.1), BCURA (2005).

Since some of the air in the bubble phase by-passes the bed, the bubble phase can be regarded as an additional air supply to the freeboard, and there is normally no need to make provision for any additional air supply direct to the freeboard as is sometimes done in incineration applications using other types of combustor.

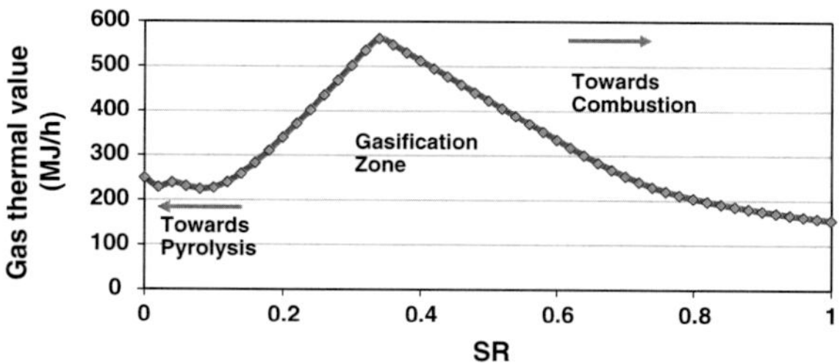

Fig. 4.4 Influence of equivalence ratio (SR) on gas thermal value from gasification of refuse-derived fuels (30 kg\h of dry municipal solid waste), Materazzi et al. (2015)

4.4.3 Number of Feed Points

Mixing of the solids in a fluidized bed is a relatively rapid process particularly in a vertical direction and radially at the top and bottom of the bed. However, if the feed rate of a fuel containing a significant proportion of volatile combustible matter, to a particular point in the bed is too great, the dense phase gas may become locally sub-stoichiometric with a consequent fall in the rate of conversion. Analogously, if the proportion of combustible matter to a particular point is too low, full combustion takes place with a consequent fall in syngas yield.

4.4.4 Bed Temperature

The bed temperature should be a minimum of 750 °C to obtain acceptable rates of conversion of the char. The char conversion rate increases with increasing temperature and hence, the higher the bed temperature, the greater the conversion efficiency, Basu and Kaushal (2009). However there must never be any danger of sintering of the bed solids and there may be other restrictions which limit the usable bed temperature.

Temperature can also influence the amount of tar formed as well as the composition of tar. Kinoshita et al. observed that the total number of detectable tar species produced from sawdust gasification decreased with increasing temperature, Kinoshita et al. (1994). Lower temperatures favoured the formation of more aromatic tar species with diversified substituent groups, while higher temperatures favoured the formation of fewer aromatic tar species without substituent groups.

The consensus seems to be that temperatures in excess of 1000 °C are necessary to destroy the refractory unsubstituted aromatics. In order to avoid melting of the inorganic phase and total defluidization of the bed, such high temperatures are not achievable in a stand-alone fluidized bed, and a separate unit (called 'thermal cracker') is normally needed, Materazzi et al. (2014).

4.4.5 Solid Fuel Feed Size

The reaction of char particles is largely, though not exclusively, a surface phenomenon. Therefore, at a constant bed temperature a greater char concentration in the bed is needed for large particles than for small ones to maintain the same burning surface area in the bed. The carbon concentration in the bed solids depends therefore on the rate of combustion and on the fuel feed size. It also depends on the fluidising velocity since the latter determines the minimum size of particle that can remain in the bed without elutriation. The carbon concentration in the bed ranges from 0.1–0.5 % w/w for solid fuels with up to 10 % inerts fed pneumatically as granules, to 2.5 % w/w or more for fuels with a high (>50 %) inerts content and low reactivity, BCURA (2005).

4.4.6 Effects of Inerts Content

The inert matter is likely to affect the fluidization characteristics. This is due to (a) different morphology, (b) different chemistry, particularly dilution of an active bed material such as olivine (c) different melting points, which will be relatively uncontrolled. The consequences of this are potentially (a) higher tar levels in the existing gas, (b) agglomeration and fusing issues in the bed, (c) changes to the fluidization characteristics which could lead to heterogeneous heat distribution (d) a change in the through-bed pressure profile.

The issue of agglomeration is potentially the most serious of these, and was raised as a concern in the literature, Öhman et al. (2000). The solutions proposed in industry indicate a requirement (a) for more frequent replacement of the bed material and (b) tighter specification of the fuel. Both of which have significant operational cost implications.

A solid inerts content in the fuel can also affect the conversion process in two other ways. Firstly, the char activity can be reduced because the inert components hinder the quantity of oxygen to the carbon in the char particles. Secondly, the higher the inerts content, the higher the bed overflow or discharge to maintain a constant bed depth—unless the inert compounds are so finely divided to leave the bed by elutriation. Carbon losses with the bed overflow depend on both the overflow rate and the carbon concentration. The latter, in turn, depends mainly on the feed size, as discussed above. Carbon losses with the bed overflow may become significant for fuels with inert contents more than 50 %. The inert stream will also remove part of the heat of reaction. If the inert material is solid this component of the heat balance is usually a minor one unless the fuel is a high ash material. However if the inert material is water, the latent and sensible heat flows may be major components-of the heat balance. Additional loss of carbon and of sensible heat in the bed solids overflow will also be incurred when additives are used for tar reforming. These losses could become significant for high tarrous fuels.

4.4.7 Bed Depth

An increase in the bed depth will give an increased residence time both for the gas and the char particles. The resulting improvement in the reaction taking place within the bed may be balanced by a decreased proportion of freeboard conversion if the freeboard temperature is allowed to decrease. However in designs where the freeboard temperature is maintained constant, or separate reforming stages are employed (such as plasma converters), then an overall improvement in gasification efficiency can be obtained.

4.4.8 Fluidizing Velocity

When a reacting char particle is reduced to that size for which the fluidising velocity equals the terminal free fall velocity then that particle will be elutriated from the bed. The fluidising velocity, therefore is probably the most important parameter affecting the in-bed thermal conversion efficiency since it determines the size of particles that will be elutriated from the bed and also the residence time of the fine particles in both the bed and the freeboard. Low velocities (below 3 m/s) result in a bubbling fluidized bed, where only a small degree of expansion ensures that bed material and char stay inside the reactor. The effect of an increase in the fluidising velocity on particle elutriation can be marked since the size of particle elutriated increases rapidly with increasing velocity. High gas velocities (5–10 m/s) cause the bed to expand throughout the entire reactor volume which results in a significant entrainment of solid particles from the reactor. For this reason, a cyclone is usually positioned directly downstream of the reactor to capture and recycle the solids fraction. This configuration is typical of circulated fluid beds. The effect of increasing the fluidizing velocity on the overall conversion efficiency may not be marked at first since increased elutriation of carbon may cause increased freeboard reactions. Nevertheless, the combined effects of an increase in the fluidizing velocity in increasing the size of particles elutriated as well as decreasing their residence time (when the material is not recirculated) will result in a significant reduction in overall conversion efficiency at high values of fluidizing velocity.

4.5 Examples of Industrial Applications

4.5.1 Fluidized-bed Combustion

Although fluidized-bed technology has a limited track record in the UK for municipal solid waste treatment, there are over 150 plants in commercial operation in Europe and Japan. This is because fluidized beds have the ability to handle waste of widely varied properties and the many advantages in controlling emissions, McLanaghan (2002).

As opposed to its direct competitor moving-grate, fluidized-bed combustion processes require the pre-sorting and processing of municipal solid waste into refuse-derived fuels. Other advantages of fluidized beds include higher combustion efficiency that is comparable to pulverised fuel-fired combustors; reduction in boiler size; low corrosion and erosion with easier ash removal; and simple operation with fast response to load fluctuations.

Since the introduction of fluid bed combustors, there has been a series of mergers and acquisitions resulting in four major market players; Alstom, Foster Wheeler, Lurgi and Kvaerner Pulping, as shown in Table 4.3. Alstom and Foster Wheeler are the largest producers of circulated fluidized-bed technology (CFB), while Kvaerner is the market leader for bubbling fluidized-bed (BFB) technology. Bharat Heavy Electricals and Energy Product of Idaho are only active in their own regions in India and North America, respectively, Koornneef et al. (2007).

The commercial capacity of the fluidized-bed combustors are influenced mainly by the cross-sectional area of the vessel. Therefore, fluidized-bed designs need to be optimised with the emphasis on outstanding engineering innovations to achieve economical vessel arrangements and reach large commercial scales. An emerging technology in this field, is the Twin-internally Circulating Fluidized-bed Furnace developed by Ebara. The technology is licensed to Lurgi and trades in Europe under ROWITEC®, which is now a well-proven process and economically a competitive option compared to moving-grate combustion.

The Madrid energy from waste incineration facility in Spain is one of the highly successful operational plants employing the ROWITEC® process and proving its operational availability in excess of 90 %, Lischke and Lehmann (2001). The plant handles 20 tonnes of waste per day, which is approximately one third of the city's

Table 4.3 Overview of fluidized-bed combustion technologies

Manufacturer	Technology	Capacity (MWe)		No. of installations	Start-up
		Min	Max		
Alstom	BFB	17	142	7	1988–99
	CFB	2	520	51	1986–2005
Babcock and wilcox	CFB	3	76	22	1982–2002
Babcock borsig	BFB	0	35	5	1982–2000
	CFB	9	120	10	1989–99
Bharat heavy electricals	BFB	5	50	18	1987–98
EPI	BFB	10	45	9	1981–93
Foster wheeler	BFB	0	117	51	1976–2002
	CFB	0	460	161	1981–2006
Kvaerner pulping	BFB	6	117	56	1985–2005
	CFB	0	240	32	1984–2002
Lurgi	CFB	9	225	35	1982–2004

Adapted from Koornneef et al. (2007)

waste and generates 25MWe of electricity that can be fed into the public grid. It also consists of sorting lines for material recycling and a composting unit.

Other successful facilities employing the ROWITEC® process include plants in Gien and Mulhouse (France), Moscow (Russia), Vienna (Austria) and the Allington plant in the UK, Fujimura and Naruse (2001). The technology has a fairly simple mechanism with no moving parts inside the furnace. It has a slanted bed floor and the air flow rate is controlled to produce a revolving sand motion. It is this mixing effect that produces a combustion performance superior to that of conventional fluidized-bed furnaces.

4.5.2 Fluidized-Bed Gasification

Large scale fluidized-bed systems have become commercial due to the successful co-firing projects, such as the Kymijärvi Power Plant at Lahti in Finland. Furthermore, fluidized beds have the advantage of extremely good mixing and high heat transfer, resulting in very uniform bed conditions and efficient reactions. Circulated fluidized-bed gasifiers, in particular, are targeted for larger scale applications (Juniper 2007), as they can be used with different fuels, require relatively compact combustion chambers and allow for good operational control. There are several leading and state-of-the-art biomass and waste fluidized-bed gasification projects across the world.

Ebara has developed a ‘new’ generation gasification technology based on its internally circulating bubbling fluidized-bed incinerator and, in 2004, had 21 process lines in commercial operation in Japan and Germany. The technology, branded TwinRec, is a state-of-the-art twin internally circulating fluidized-bed gasifier. It is designed with ash vitrification technology for material recycling, energy recovery and detoxification of waste in an integrated and economical process. The gasifier is a revolving fluidized bed, which gasifies waste and produces heat that is used to raise the temperature in the next-stage slag combustion furnace. Due to the high temperatures inside the furnace, dioxins are decomposed and the ash is vitrified and recycled as stable glass granulates. Aomori is the largest gasification and slagging combustion system in Japan, with a capacity of 450 tonne per day and a power output of 17.8 MWe using a steam turbine, Selinger and Steiner (2004).

Enerkem’s Biosyn gasification process is based on a bubbling fluidized-bed gasifier that operates at 700–900 °C and up to 1.6 MPa. The process proved the technical feasibility of gasifying biomass from forest and agricultural residues, as well as refuse-derived fuels, rubber resides and sludge, Yassin (2008). The technology is available in the UK and Ireland under license by Novera Energy Europe. The Novera/Enerkem gasification technology is built cost-effectively at a smaller scale than combustion processes so it complies well with the proximity principal for waste disposal. The process has a low emission profile and is easily operable well inside the emission limits set under the Waste Incineration Directive. In late 2006, Novera Energy has signed a contract with Defra to build a gasification plant at the Ford plant in Dagenham in partnership with East London Waste Authority, Shanks and the Ford Motor Company. The plant will process 90,000–100,000 tpa of RDF supplied from the nearby Shanks MBT (mechanical biological treatment) plant at Frog Island. It will provide Ford with 8–10 MWe of electricity, which is equivalent to approximately £4 million per annum worth of electricity purchased from the national grid, while East London Waste Authority will benefit through the Landfill Allowance Trading Scheme, Yassin (2008).

Foster Wheeler has been supplying fluidized-bed gasifier systems for many years. The Kymijärvi Power Plant at Lahti in Finland is one of the most successful commercial demonstration plants coupling gasification with co-firing. The plant is a pulverised coal fired steam plant that generates up to 167 MWe of electricity and up to 240 MWth of district heat. It uses a circulated fluidized-bed gasifier to produce a low calorific product gas, which is combusted in the coal-fired boiler, thus replacing about 30 % of the coal. The gasifier uses biofuels, such as saw dust, wood residues and recycled fuels comprising of cardboard, paper and plastics, Spliethoff (2001). In addition, FW has contributed to the construction of the first complete Integrated Gasification Combined Cycle power plant at Värnamo in Sweden. The demonstration plant employed a pressurised air-blown circulated fluidized-bed gasifier operating at 950–1000 °C and 2 MPa, Stahl and Neergaard (1998). It fed about 6 MWe of electricity to the grid and 9 MWth of heat to the district heating network of the city of Värnamo.

FERCO has acquired the SilvaGas process from Battelle, who started developing this gasification process in 1977. The process uses forest residue, MSW, agricultural waste and energy crops and converts them into a syngas. The SilvaGas process consists of two interconnected atmospheric pressure circulated fluidized-bed reactors for steam gasification in one reactor, and a residual char oxidation with air in the second one, with solids exchange between the two reactors. The first commercial scale biomass gasification demonstration plant based on the SilvaGas process was built at the McNeil Power Station in Burlington, Vermont, USA. The syngas was used as a co-fired fuel in the existing McNeil power boilers and in a combined cycle with a gas turbine power generation system, Paisley et al. (1997).

The Gas Technology Institute, through its predecessor organisations (the Institute of Gas Technology and Gas Research Institute), has originally developed the air-blown Renugas technology for Integrated Gasification Combined Cycle gasification applications. The technology is based on a single stage pressurised bubbling fluidized-bed gasifier, with a deep bed of inert solids, which is also capable of producing a hydrogen-rich fuel. A 15 MWth pilot plant was commissioned in 1993 in Tampere (Finland) by Carbona, who licensed the Renugas technology from the Gas Technology Institute. The plant has operated for more than 2000 h on paper mill wastes, straw and coal mixtures, alfalfa stems and a variety of wood fuels, Arrieta et al. (1999). In 2004, Carbona has signed a contract to build a biomass combined heat and power gasification plant in Skive (Denmark). The plant will produce 5.5 MWe of electricity using gas engines and 11.5 MWth of district heat for the town of Skive, Babu (2005).

The Lurgi circulated fluidized-bed gasifiers operate at near atmospheric pressure and are well suited for capacities up to 30 t/h of feedstock. The gasification plants in Pöls (Austria) and Rüdersdorf (Germany) were designed and constructed for use in cement industry. The Bioelecttrica project in Italy uses an atmospheric circulated fluidized-bed gasifier integrated with a combined cycle of a 10.9 MWe gas turbine and a heat recovery steam generator of 5 MWe. The fuel used is a mixture of wood chips, as well as forest and agricultural residues. The project was aimed at the demonstration of the technical and economic feasibility of power generation from biomass using Integrated Gasification Combined Cycle. In 2000, Lurgi has contributed to the construction of the 85 MWth circulated fluidized-bed wood gasification process at the AMER9 power plant in the Netherlands. The syngas from the gasification process is co-fired in a pulverised coal combustor unit replacing 70,000 tpa of coal, Willeboer (1998).

The gasification process developed by TPS is based on an atmospheric circulated fluidized-bed gasifier operating at 850–900 °C and is coupled to a dolomite-containing tar-cracking vessel. A pilot-scale refuse-derived fuel gasification plant was commissioned in Grève-in-Chianti (Italy) in 1992. The plant processes 200 tonnes of waste refuse-derived fuel per day, which is fed into two air-blown circulated fluidized-bed gasifiers, each with a fuel capacity of 15 MWth. The syngas is used in a steam boiler to drive a 6.7 MWe steam turbine. In the UK, the gasification technology of TPS was installed in a wood-fuelled Integrated Gasification Combined Cycle plant at ARBRE, Eggborough in Yorkshire (shut down in 2002). The syngas from the process was compressed and combusted in a combined cycle gas turbine to produce 8 MWe of electricity. In Brazil, there are two projects based on the TPS technology, which aim to demonstrate the commercial viability of biomass fuelled Integrated Gasification Combined Cycle using gas turbines. The first is a 32 MWe plant that utilises wood as a feedstock, while the second plant uses sugar cane baggasse and cane trash, with the intention of integrating the biomass Integrated Gasification Combined Cycle system into a typical sugar mill, Yassin (2008).

The Austrian Institute of Chemical Engineering at the Technical University of Vienna (TUV) and AE Energietechnik have developed a novel fluidized-bed gasifier reactor producing a product gas with a high calorific value of up to 15 MJ/Nm^3. The gasification process is based on fast internal circulating fluidized bed and consists of a gasification zone fluidized with steam and a combustion zone fluidized with air. The circulating bed material acts as heat carrier from the combustion to the gasification zone, Vreugdenhil (2010). A demonstration combined heat and power plant located in Güssing, Austria, applies this technology and it produces 4.5 MWth for district heating and 2 MWe from an 8 MWth fuel input. The plant was commissioned in 2001 and uses wood chips and residues from industry as feedstock.

Table 4.4 Summary of main biomass and waste fluidized-bed gasification projects

Gasification type	Technology developers
Heat gasifiers (syngas combustion)	
Pöls, Austria	27 MWth CFB, Lurgi
Rüdersdorf, Germany	100 MWth CFB, Lurgi
Co-firing gasifiers	
Amer, Netherlands	85 MWth CFB, Lurgi
Burlington, USA	50 MWe CFB, Battelle
Lahti, Finland	40–70 MWth CFB, FW
Ruien, Belgium	50 MWth CFB, FW
Zeltweg, Austria	10 MWth ACFB, AEE
IGCC plants	
ARBRE, UK	8 MWe CFB, TPS
Grève-in-Chianti, Italy	6.7 MWe CFB, TPS
Pisa, Italy	12 MWe CFB, Lurgi
Värnamo, Sweden	18 MW PCFB, FW
CFB gasifiers with gas engine	
Güssing, Austria	8 MWth FICFB, AICE
Skive, Denmark	11.5 MWth PBFB, Carbona

PBFB = Pressurised bubbling fluid bed
AICE = Austrian Institute of Chemical Engineering, TUV

Finally, the main biomass and waste fluidized-bed gasification projects covered in this section are summarised according to their configuration in Table 4.4.

4.5.3 *Pyrolysis*

Pyrolysis is a less proven technology when compared to gasification and combustion and has a limited track record in the UK on the treatment of municipal solid waste. Whilst established pyrolysis technologies for the treatment of certain specific waste streams exist, it is only in recent years that pyrolysis has been commercially applied to the treatment of municipal solid waste. Nonetheless, the liquid bio-oil has a considerable advantage of being storable and transportable, as well as the potential to supply a number of valuable chemicals. In this respect, it offers a unique advantage and should be considered complementary to the other thermal conversion processes. Although the best reactor configuration is not yet established, fluidized-bed technology, as for gasification, is one of the most efficient and economic technologies of actualising fast pyrolysis as it offers high heating rate, rapid devolatilisation and convenient char collection and re-utilisation. Ensyn and Dynamotive are major developers of fluidized-bed pyrolysis technologies.

Ensyn has been producing commercial quantities of bio-oil from its Rapid Thermal Process (RTP™), which uses a circulated fluidized-bed reactor, since 1989. The RTP™ produces liquid bio-oil, gas and charcoal, which can be sold as fuel. Ensyn has developed natural chemical products from the liquid that have a much higher value. These include food flavourings and other products that can replace petroleum-based chemicals. In addition, the charcoal by-product is easily and economically upgraded to a higher value carbon product. The RTP™ is characterised by a very rapid heat addition and very short processing times of typically less than one second at moderate temperatures and atmospheric pressure. The 70 tonne per day RTP™ facility in Wisconsin produces a number of food, natural chemical and liquid bio-fuel products and operates with an availability exceeding 95 %. Ensyn has supplied a 650 kg/h unit to ENEL in Italy and a 350 kg/h unit to Fortum in Finland, Yassin (2008).

Dynamotive owns the rights for its BioTherm™ process, which incorporates a bubbling fluidized-bed pyrolyser, originally developed by Resources Transforms International. The process produces high quality bio-oil, char and non-condensable gases, which are recycled to supply 75 % of the energy required by the process. The bio-oil can be used directly in gas turbines or diesel engines for power generation. The company is also developing a range of derivative bio-oil products including blended fuels, slow release fertilisers and speciality chemicals, such as BioLime®, a reagent used to control SOx and NOx emissions in coal combustion systems. In 2005, Dynamotive has entered the commercialisation phase with the launch of its 2.5 MWe combined heat and power facility in West Lorne, Ontario (Canada). This is the first bio-oil combined heat and power facility and is capable of processing 100 tonnes per day of bio-fuel, mainly wood, and incorporating a 2.5 MWe gas engine, Yassin (2008).

4.5.4 Fluid Bed Plasma Treatment

The use of plasma has increasingly been applied with pure waste treatment for its ability to completely decompose the input waste material into a tar-free synthetic gas and an inert, environmentally stable, vitreous material known as slag. Advanced thermal treatments of low grade fuels that utilise plasma, are demonstrated to produce a clean syngas suitable for energy production as a by-product of the vitrification of hazardous waste streams such as organic pollutants or oily sludges. These plants produce a tar free syngas but a relatively high degree of input electrical energy is required, especially for large scale plants where massive quantities of waste need to be treated. A possible solution is seen in focussing the application of the plasma to what it does best—the vitrification of solid residues (ash) and the cracking of tars and long-chain (and aromatic) hydrocarbons. The process of primary conversion itself (either gasification or pyrolysis) could be achieved most efficiently and at very significant scale by employing a conventional fluid bed reactor upstream of the discrete plasma step. Overall, the energy balance of combining these process steps in this way is significantly better than by direct plasma application. The carbon conversion efficiency is very high, plasma parasitic load is kept low and heat is recovered directly from syngas cooling, Materazzi et al. (2013).

Advanced Plasma Power (UK) developed a two stage process (the Gasplasma process) which combines fluid bed gasification with plasma technology (Fig. 4.5), to produce a clean low inert containing synthesis gas, for use directly in power generation in gas engines/gas turbines or as a precursor for waste-to-gas and waste-to-liquid applications whilst also recovering significant quantities of heat.

The first stage is a bubbling fluidized-bed gasifier operated in temperature range between 650 and 800 °C, where the intense gas/solids contacting ensures the high heat transfer and reaction rates required to efficiently gasify the waste fuel. The second stage is a direct current plasma converter that 'polishes' the producer gas by organic contaminants and collects the inorganic fraction in a molten (and inert) slag. This slag is formed by the melting ash particles that continuously separate from the gas stream. Additional solids (oversize material) from the BFB bottom ash may be fed into the furnace to form more slag. Unlike some other gasification technologies, there is no need of intermediate fuel gas cleanup between the gasifier and the ash melting plasma converter. The plasma power is controlled to provide a uniform syngas temperature ($\sim$1200 °C) and destruction of the residual tars and chars contained within the crude syngas. Downstream of the plasma converter, the syngas can be directed straight to a solid oxide fuel cell stack for power generation, or cooled to around 200 °C in a steam boiler prior to cleaning treatment to remove any residual particulates and acid gas contaminants. The refined gas can be then used for power generation (gas engines or gas turbines), for conversion to a liquid fuel, or used as a chemical precursor. The Gasplasma process delivers energy conversion rates of 90 % in terms of syngas production; the net exportable power generation efficiency for a commercial scale plant is significantly in excess of 25 %. Based on 100,000 tonnes per annum input of a typical refused derived fuel, a Gasplasma facility generates in excess of 20 MW of electrical power, Taylor et al. (2013).

The main advantage of coupling fluidized-bed gasifier and plasma technologies is that the oxidant addition rate and power input in the two-stage process can be controlled independently while, unlike single stage fluidized-bed gasifiers, the

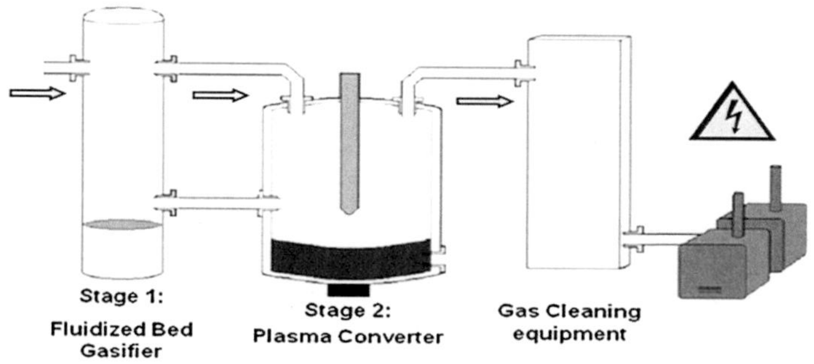

Fig. 4.5 Gasplasma schematic

gasification stability is not dependent on the gas evolved from the fuel itself. The energy input from plasma is readily controllable and (unlike oxidizing systems) is independent of the process chemistry, generating low off-gas volumes, reducing the size, complexity and the associated capital and operating cost of the downstream gas cleaning equipment and rendering the process flexible to changes in the refused derived fuel characteristics, typical of waste materials, Bosman et al. (2013).

Tar and ashes treatment from fluid bed reactors by plasma methods is one of the most concern topics for current scientific research and numerous treatment methods regularly emerge from the scientific community and are reported to be very effective in tar reduction and ash disposal but still need to be optimized to be economically viable and used industrially.

4.6 Conclusions

Thermal treatment processes including combustion, gasification and pyrolysis recover energy from waste in the form of heat and/or power. The heat can be used for district heating and the power can be easily distributed and sold via the national grid. Gasification and pyrolysis have the added advantage of producing a syngas that can be burned in conventional steam turbines or utilised in high efficiency gas engines and turbines. The syngas can also be further processed via gas synthesis to produce speciality chemicals and liquid fuels.

Fluidized-bed technologies offer alternative and reliable options to other Waste-to-Energy technologies because of their ability to handle waste of widely varied properties and the many advantages in controlling emissions. Although the technology has a limited track record in the UK for municipal solid waste treatment, there are over 200 plants in commercial operation in Europe and Japan.

The general conclusion drawn from the literature and experimental tests is that it is technically possible to convert a given material satisfactorily in a fluidized-bed reactor provided the following conditions are met:

- The material must be capable of being fed in such a way that it is quickly and evenly dispersed throughout the bed and remains there for a sufficient time for its conversion to be completed to an acceptable degree.
- The bed temperature must be such that there is no tendency for softening and melting (and consequent agglomeration) of any of the solid components in the fluidized bed.
- The fluidized bed must be maintained in a “well fluidized” state at all times. This condition implies that the fuel and its conversion products must be maintained in conditions such that no uncontrolled agglomeration of bed solids occurs.
- An appropriate provision for the removal of oversized bed material must be made for the treatment of fuels containing a significant proportion of oversize inert solid material (e.g. municipal solid waste).

Experimental testing will often be required to determine the optimum operating conditions for a particular fuel and application that comply with the above requirements. It should also be borne in mind that although it may be possible to treat certain materials using a stand-alone fluidized-bed reactor, it does not necessarily follow that it will be the best solution to do so. Often, complementary reforming steps in a multiple stage process can overcome most of the limitations of stand-alone reactors for use in waste thermal treatments.

References

Anthony EJ (1995) Fluidized bed combustion of alternative solid fuels: status, success and problems of the technology. Prog Energy Combust Sci 21:239–268

Arrieta FRP, Sanchez CG (1999) BIG-GT and CEST technologies for sugar cane mill. Thermodynamic and economic assessments. In: Proceedings of the 2nd olle lindström symposium on renewable energy, Bioenergy, Stockholm, Sweden, pp. 184–189, June 9–11 1999

Babu S (2005) Biomass gasification for hydrogen production—process description and research needs. Technology report. Task 33: Thermal gasification of biomass. IEA Bioenergy, Dublin, Ireland

Basu P, Kaushal P (2009) Modeling of pyrolysis and gasification of biomass in fluidized beds: a review. Chem Prod Process Model 4(1)

Bosmans A, Vanderreydt I, Geysen D, Helsen L (2013) The crucial role of waste-to-energy technologies in enhanced landfill mining: a technology review. J Clean Prod 55:10–23

British Coal Utilisation Research Association (BCURA) (2005) The Combustion Systems Ltd., (CSL). Fluidized combustion process design manual

Chirone R, Massimilla L, Salatino P (1991) Comminution of carbons in fluidized bed combustion. Prog Energy Combust Sci 17:297–326

Defra (2013) Advanced thermal treatment of municipal solid waste. Department for Environment, Food and Rural Affairs (Defra), London

Doug Orr DM (2000) A comparison of gasification and incineration of hazardous waste. U.S. Department of Energy, Morgantown

Fujimura H, Naruse K (2001) Fluidized-bed gasification and slagging combustion system. In: Proceedings of the IT3 conference, Philadelphia, USA, 14–18 May 2001

Gómez-Barea A, Leckner B (2010) Modeling of biomass gasification in fluidized bed. Prog Energy Combust Sci 36(4):444–509

Gomez-Barea A, Nilsson S, Barrero FV, Campoy M (2010) Devolatilization of wood and wastes in fluidized bed. Fuel Process Technol 91(11):1624–1633

Grace JR, Clift R (1974) On the two-phase theory of fluidization. Chem Eng Sci 29(2):327–334

Juniper (2007) Advanced conversion technology (Gasification) for biomass projects. Juniper, Uley, UK

Kersten SRA, Wang X, Prins W, Van Swaaij W P M (2005) Biomass pyrolysis in a fluidized bed reactor. part 1: literature review and model simulations. Ind Eng Chem Res 44(23):8773–8785

Kinoshita CM, Wang Y, Zhou J (1994) Tar formation under different biomass gasification conditions. J Anal Appl Pyrol 29(2):169181

Koornneef J, Junginger M, Faaij A (2007) Development of fluidized bed combustion—an overview of trends, performance and cost. Prog Energy Combust Sci 33:19–55

Lischke G, Lehmann E (2001) Madrid integrated waste processing facility—sorting, composting, ROWITEC® fluidized bed incineration. In: IMECHE conference transactions, Sixth international conference on engineering for profit from waste, Wiley, New York, USA, 13–30

Materazzi M, Lettieri P, Mazzei L, Taylor R, Chapman C (2013) Thermodynamic modelling and evaluation of a two-stage thermal process for waste gasification. Fuel 108:356–369

Materazzi M, Lettieri P, Mazzei L, Taylor R, Chapman C (2014) Tar evolution in a two stage fluid bed-plasma gasification process for waste valorization. Fuel Process Technol 128:146–157

Materazzi M, Lettieri P, Mazzei L, Taylor R, Chapman C (2015) Fate and behaviour of inorganic constituents of RDF in a two stage fluid bed-plasma gasification plant. Fuel 150:473–485

McLanaghan SRB (2002) Delivering the landfill directive: the role of new and emerging technologies. A report for the strategy unit, AIIE, Penrith, UK

Miles TR, Miles TR Jr, Baxter LL, Bryers RW, Jenkins BM, Oden LL (1996) Boiler deposits from firing biomass fuels. Biomass Bioenergy 10(2–3):125–138

Öhman M, Nordin A, Skrivars B-J, Backman R, Hupa M (2000) Bed agglomeration characteristics during fluidized bed combustion of biomass. Energy Fuels 14:169–178

Oka S (2004) Fluidized bed combustion. CRC Press, Marcel Dekker, Basel

Paisley MA, Farris G, Slack W, Irving J (1997) Commercial development of the Battelle/Ferco biomass gasification—initial operation of the McNeil gasifier. In: Overend RP, Chornet E (eds) Making a business from biomass in energy, environment, chemicals, fibers and materials. Proceedings of the third biomass conference of the Americas. Elsevier, New York, USA, 579–588

Ross IB, Davidson JF (1981) The combustion of carbon particles in a fluidized bed. Trans Inst chem Eng 59:108–114

Saxena SC, Jotshi CK (1994) Fluidized-bed incineration of waste materials. Prog Energy Combust Sci 20:281–324

Scala F (2013) Fluidized bed technologies for near-zero emission combustion and gasification. Woodhead Publishing Ltd, Cambidge

Selinger A, Steiner C (2004) Waste gasification in practice: TwinRec fluidized bed gasification and ash melting—review of four years of commercial operation. In: Proceedings of the IT3 conference, Phoenix, USA, 10–14 May 2004

Spliethoff H (2001) Status of biomass gasification for power production. IFRF Combust J

Ståhl K, Neergaard M (1998) IGCC power plant for biomass utilisation, värnamo, sweden. Biomass Bioenergy 15(3):205–211

Taylor R, Ray R, Chapman C (2013) Advanced thermal treatment of auto shredder residue and refuse derived fuel. Fuel 106:401–409

Vreugdenhil BJ (2010) Alkali distribution for low temperature gasification. ECN-E–10-103

Willeboer W (1998) Amer demolition wood gasification project. Biomass Bioenergy 15:245–249

Williamson J, West SS, Laughlin MK (1994) The behaviour of bed material during fluidized bed gasification: the effects of mineral matter interactions. Fuel 73(7):1039–1046

Yassin L (2008) Appropriate scales for energy recovery from urban waste using fluidized bed technology. Ph.D. thesis, University College London

Chapter 5
Effect of Process Conditions on Fluidization

Abstract Previous chapters have illustrated the variety of fluidized-bed industrial applications and the importance of the process conditions on their operation. This chapter reviews experimental and theoretical studies on the influence of process conditions (temperature, pressure, presence of liquid, fines and fines size distribution) on the fluidization quality of gas-solid fluidized-bed reactors. The chapter begins with an overview of the effect of process conditions on fluidization highlighting the role of the hydrodynamic and interparticle forces on fluidized-bed behaviour. A brief review of the interparticle forces is reported to explain the foundation for the understanding of the factors responsible for the changes in fluidization at process conditions. Hence, the chapter discusses specifically the effect of temperature, pressure and other special conditions in the fluid bed, at minimum fluidization conditions, in the expanded fluid bed and at minimum bubbling conditions, showing how correlations and models established at ambient temperature and pressure may lead to misleading predictions at super- ambient conditions.

5.1 Introduction

A great deal of research has been carried out at ambient conditions with special attention being paid to evaluate the effect of physical properties of the particles on the enhancement of gas-solid contact and, as a consequence, chemical conversion; see Rowe et al. (1978) and Grace and Sun (1991). When we are considering the stability of a suspension, particle size and particle size distribution become important. For example, a suspension of 1 μm particles in air may remain stable for many minutes, whereas 100 μm particles will settle out in seconds. Similarly, flow rates of particles from hoppers, standpipes and most other aspects of particle behaviour depend on particle size. It is also known that addition of fine particles to a powder of coarser particles tends to improve its fluidization characteristics.

We dedicate this chapter to the late Dr. David Newton (formerly Head of the Fluidization Group at BP Chemicals Sunbury), a close colleague and friend, who contributed significantly to the work described herein.

J.G. Yates and P. Lettieri, *Fluidized-Bed Reactors: Processes and Operating Conditions*, Particle Technology Series 26,
DOI 10.1007/978-3-319-39593-7_5

It is on the basis of a variety of small scale tests developed at ambient temperature and pressure that fluid-dynamic models and correlations have been established and have been used for design criteria and performance predictions for fluid bed units working at high temperature and pressure. For a long time, the influence of the operative conditions on the fluid-dynamic characteristics of the system has been considered by simply accounting for the variations of the gas properties, namely its density and viscosity. However, extrapolating results and relationships from those developed at ambient conditions is reliable only when the hydrodynamic forces (HDFs) dominate the fluidization behaviour. Overlooking possible modifications induced by temperature and pressure to the structure of the fluidized bed, which can cause drastic changes in the fluidization behaviour and stability of the powders between ambient conditions and at high temperatures and pressures, is likely to lead to a misleading prediction of the fluid-bed performance and thus to errors in evaluating heat and mass transfer phenomena. A "reliable prediction" of the fluidization behaviour at unit operational conditions is of major importance, given that many of the industrial plants exploiting fluidization technology have been designed for operations run at thermal levels and pressure well above the ambient conditions, as described in the previous chapters.

Given the relevance of its applications, research on the influence of temperature and pressure on fluidization has been gaining interest, but findings are still controversial, as reported by Knowlton (1992) and Yates (1996) in their reviews on the subject. The positive effect of increased pressure in a fluidized bed is known to enhance bed-to-surface heat transfer coefficients in beds of Geldart Group A powders because of the suppression of bubbling, while in beds of Group B materials the enhancement is through an increase in the gas convective component of the transfer coefficient (see Sect. 5.3.3). Increased temperature can be responsible for modifications in the structure of fluidized beds causing in turn dramatic changes in the fluidization behaviour. A satisfactory understanding of the phenomena which are responsible for such changes has not yet been achieved. Much of the controversy still holds because the relative importance of the interparticle forces (IPFs) and hydrodynamic forces (HDFs) on the flow behaviour of the particles remains undefined.

Most of the disagreement on the relative role of HDFs and IPFs on the fluidizability of powders lies in the uncertain nature on the IPFs involved and in the difficulty of measuring them directly. Seville and Clift (1984) approached this problem introducing IPFs in a controlled manner and monitoring changes in the fluidization behaviour. Lettieri (1999) showed how the combined effect of temperature and presence of liquid can enhance the role of IPFs causing changes in the fluidization behaviour of industrial powders.

When trying to describe the fluidization of different materials the nature of the forces acting between adjacent particles becomes of major importance. It is well-known that finely divided Group C powders are very difficult to fluidize. The commonly accepted reason for this behaviour is the dominance of surface forces. The ratio of the surface forces to body forces increases with diminishing particle size. Hence, the fluidization of very fine materials, belonging to Group C is dominated by the interparticle forces, which are greater than those transmitted to the particles by the fluidizing gas, Baerns (1966).

On the other hand interparticle forces are considered negligible when studying the fluidization behaviour of Group B and Group D powders. It is well established in the literature that interparticle forces also exist in Group A powders, although their importance as compared to body forces is not yet unequivocally defined. This is mainly due to the difficulty in recognizing the nature of the interparticle forces involved and, therefore, to quantify their effect on the fluidization behaviour.

The debate on the role of the IPFs and HDFs on the stability of Group A powders still divides into two groups the scientific world working on this matter. The physical origin of the stable behaviour of Group A powders has been studied theoretically, and two different approaches have been taken, one being based on the contention that bed stability is dominated by the hydrodynamic forces, see Foscolo and Gibilaro (1984), and the other that the interparticle forces are the controlling factor, see Mutsers and Rietema (1977). The physical origin of the stability of Group A powders has also been studied at an experimental level. Abrahamsen and Geldart (1980) and later Xie and Geldart (1995) investigated the stable behaviour of Group A materials by using a measurable parameter, the u_{mb}/u_{mf} ratio, which they defined as capable of predicting the aeratability of the powders. Xie and Geldart (1995) used the u_{mb}/u_{mf} ratio to correlate their experimental results obtained on the entire range of Group A materials, changing gas adsorption and operational conditions. They concluded that this parameter reflects both the effects of interparticle forces and hydrodynamic forces on the fluidization behaviour of fine powders. However, the usefulness of the u_{mb}/u_{mf} ratio as a discriminating test between low and high temperature was critically assessed by Newton et al. (1996) on the basis of experimental results obtained from fluidization at high temperature of some FCC catalysts.

Various other authors studied experimentally the stability of Group A powders with increasing temperature. Much debate on the interpretation of the results and the relative importance of the IPFs and HDFs is still in progress, mainly due to the difficulty of recognizing the nature of the interparticle forces involved and therefore of quantifying their effect on the fluidization behaviour. It is therefore necessary at this point to review the types and nature of the interparticle forces which might be encountered.

5.2 Interparticle Forces

Particle-particle contacting can be the result of different mechanisms of adhesion, the ones discussed in this chapter are shown in Table 5.1. An extensive review on the subject is reported in Israelachvili (1991).

Initially, the interparticle forces which arise without material bridge are discussed. A review on the capillary forces follows. Finally, the effect of temperature on the properties of the particle surface is discussed. A review on the formation of solid and sintered bridges is also presented.

Table 5.1 Mechanism of adhesion

Without material bridges	With material bridges
Van der Waals forces	Capillary forces
Electrostatic forces	Solid bridges
Magnetic forces	– Sintering
Hydrogen bonding	

5.2.1 Van Der Waals Forces

Electrostatic, capillary and van der Waals forces are said to be the most important to fluidized beds of fine powders. These forces depend on the particle size and the interparticle separation, usually becoming stronger with decreasing particle size and particle-particle separation. Other factors such as particle shape, surface roughness, gas humidity, moisture content and contamination also play a role. These factors can be affected by process conditions, for example high temperature.

Molecular or van der Waals forces arise from random motion of the electrons in the surface molecules. They are comprised of three types:

- Forces between polar molecules
- Forces between molecules polarised by fields of other molecules
- Forces of dispersion between non-polar molecules, due to the local polarization produced in molecules by the random fluctuation of electrons.

Intermolecular and interparticle forces are very different. The intermolecular forces decay with increasing molecular separation, z_o, as z_o^{-7}, whereas the interparticle forces as z_o^{-2}. In order to scale up the van der Waals forces to bodies having sizes larger than the molecular dimension, the Hamaker theory (Hamaker 1937) can be used. This assumes that the interaction energies between the isolated molecule and all the molecules in the large body are additive and non-interacting. Thus, the net energy can be found by integrating the molecular interactions over the entire body. The attraction force, F_a, for two perfectly spherical and rigid particles having diameters d_1 and d_2 at a separation distance a is:

$$F_a = \frac{AR}{12z_o^2} \tag{5.1}$$

where $R = d_1d_2/(d_1 + d_2)$, A is the Hamaker (materials-related) constant and z_o is the surface separation, which takes a minimum value of the order of the intermolecular spacing (generally assumed to be 4 Å). Values for the constant A can be found in Israelachvili (1991). The range of values for the Hamaker constant are quite small. For most solids interacting across vacuum or air $A \sim 4–40 \times 10^{-20}$ J.

Rietema et al. (1993) calculated the minimum value for the parameter a taking into account a repulsive force as well as attractive and using a net force $F_{attractive}$-$F_{repulsive}$. In this way they evaluated a smaller value for the minimum surface separation of 2.23 Å. In the light of this calculation they estimated also the cohesion force due to van der Waals forces for two perfectly spherical and rigid particles having diameters and density of a typical Group A material. This was several order of magnitude greater than the gravitational force. Rietema et al. (1993) also elaborated a fairly complicated model to account for particle deformation when evaluating the cohesive force between particles. Rietema and Piepers (1990) and Musters and Rietema (1977) interpreted the role of the van del Waals forces as the origin of the interparticle forces. They assumed that van del Waals forces are the controlling factor in the stable behaviour of group A powders, as opposed to the theory developed by Foscolo and Gibilaro (1987), already introduced in Chap. 1, according to which hydrodynamic forces dominate the transition from particulate to bubbling fluidization.

Massimilla and Donsi' (1976) also used Eq. 5.1 to calculate the van der Waals attractive forces between rigid particles. They used the following binomial formula to account for particle deformation:

$$F_a = \frac{A}{6a^2}\left(1 + \frac{A}{6\pi a^3 H}\right)R \tag{5.2}$$

where H is the hardness of the softer of the bodies in contact, which they quoted to be 10^7 N/m^2 for FCC catalysts. The cohesion force increased by several hundred times when using Eq. 5.2. Massimilla and Donsi' (1976) stated that the second term in Eq. 5.2 is negligible for materials with hardness greater than 10^7 N/m^2.

The comparison between fluidization behaviour of Group A powders and the magnitude of the cohesive force obtained using Eq. (5.1) or (5.2), led Massimilla and Donsi' (1976) to investigate the particles' surface, in order to establish correct values for local radii of curvature R to enter into the equations. They observed the presence of surface asperities in the form of sub-particles, and were a common characteristic of all materials analyzed. Massimilla and Donsi' (1976) stated that such asperities become the sites at which contact takes place. Thus, the contact forces between solids are smaller by orders of magnitude according to the ratio between sub-particles diameter and particle size. By accounting for surface irregularities the cohesive force can be reduced by about two orders of magnitude. However, this still leaves the cohesive forces greater than the particle weight. Massimilla and Donsi' (1976) showed that for particles having diameters above 40 μm, the cohesive forces remain constant, while gravity forces increase with the cube of the particle diameter. This confirms the well-known reduction of the influence of the interparticle forces as the diameter of the particles increases.

In conclusion it can be said that whatever method is used it results in an overestimation of the attraction force between two particles, particularly if the deformation of the particles is accounted for. If the cohesive forces were this large the natural state of a Group A powder would be paste-like and it would never be fluidizable.

5.2.2 Electrostatic Forces

Particle adhesion due to static electricity is caused by the motion of electric charges on the surface of the particles at contact. This leads to the formation of an electric double layer surrounding the charged particles, in which positively charged elements prevail on one side and negatively charged ones on the other. Electrostatic forces depend on a number of variables difficult to evaluate, such as particle local geometry, surface roughness, presence of impurities, humidity and moisture in the molecular structure.

A fluidized bed is, by its very nature, a place where continuous contact and separation of solid particles occur, as well as the friction of the particles against others, and against the walls of the fluidized-bed container. Such circumstances should favour charge generation during fluidization, which may represent a potential safety hazard.

Boland and Geldart (1971) were amongst the first authors to contribute to the understanding of electrostatic charging in fluidized beds. They found that most of the particle-particle charging in the bed was associated with the passage of bubbles. They measured opposite sign voltages at the nose and wake regions of the bubbles, a phenomenon not entirely understood, but which led to the suggestion that a different mechanism of charge transfer takes place at the nose and wake region of the bubble. Frictional and kinetic effects may be more important in the wake region where particle motion is more intense, and consequently particle charging higher. It was also thought that a difference in voidage between the nose and the wake region could be the cause of the change in resistance, resulting in a different mechanism of charging. Electrostatic forces are difficult to control; however, by increasing the relative humidity of the fluidizing gas and the conductivity of the particles' surfaces it is possible to reduce the electrical resistance of the particles.

5.2.3 Magnetic Forces

A comprehensive survey on the effect of magnetic forces on fluidized beds was reported by Siegell (1989). Siegell reported that in the bubbling regime magnetic fields, with the smallest gradients along the height of the bed, produce a more uniform fluidization. Measurements of pressure fluctuations in the bed were greatly reduced with increasing magnetic field strength. More uniform porosity distribution in the bed was also reported as an effect of the bed magnetization, the latter causing though a decrease in the heat transfer coefficient. Given the stabilizing effect of magnetic fields, magnetized fluidized beds have been used to improve different industrial processes.

5.2.4 Capillary Forces

When a powder is in equilibrium with a dry atmosphere, at ambient conditions, the electrostatic or magnetic forces may be the only forces to consider at the contact point with another particle. If the humidity of the atmosphere is increased, then capillary forces may become an important component of the interparticle forces. At low humidity, capillary forces are caused by adsorption of water vapours on the surface of the particles. In this case the adhesion force between two particles depends on them coming close enough together for the adsorbed layers to overlap. As the relative humidity approaches saturation, then condensation occurs, causing the thickness of the adsorbed liquid layers to increase and generate more stable liquid bridges at the contact point between particles.

The mechanism of particle agglomeration due to liquid bridges has been widely studied given its importance in various industries. For example, it is beneficial in the process of granulation, which is extensively applied in pharmaceutical, mineral and fertilizer industries. However, it can also be deleterious causing serious problems in the handling of sticky particulate materials. In agglomeration processes the capillary forces can become so strong that fluidization can be lost completely, a phenomenon known as “wet quenching”. On the other hand, the liquid bridges may subsequently evaporate, leaving the particles permanently agglomerated in solid bridges, and give place to a phenomenon called “dry quenching”.

D’Amore et al. (1979) reported on the influence of moisture on the fluidization characteristics of non-porous and porous materials. They emphasized that particle porosity is the property which affects the ability of the materials to retain water without losing their fluidizability characteristics. Seville and Clift (1984) reported on the effect of liquid loading on the fluidization of Group B materials. They observed changes in the fluidization behaviour, which shifted through Group A to C, upon the increasing addition of liquids and the corresponding increase of the IPFs generated. Tardos et al. (1985) also studied the destabilization of fluidized beds due to agglomeration. They found that the limiting velocity at which the bed could be still fluidized was dependent on the amount of liquid added as well as the bed and fluid properties.

The approach taken to model the agglomeration process has been to scale up forces between pairs of particles to systems of multi-particles such as fluidized beds. Two different approaches have been developed to model the behaviour of wet agglomerates, one based on the assumption that the dynamic forces (dominated by viscosity) are the controlling factor, the other that the static forces (dominated by surface tension) are more important. Ennis et al. (1991) in their work on granulation phenomena between wet particles neglect static, consider that the energy loss during collision of two particles is due to the viscous dissipation in the liquid layer. They introduced a viscous Stokes number to predict the minimum velocity required for two coated spherical particles to rebound:

$$\mathrm{St}^* = \frac{2\,\mathrm{m}\,\mathrm{v_o}}{3\,\pi\,\mu_l\,\mathrm{R}^2} = \frac{8\,\rho\,\mathrm{v_o}\,\mathrm{R}}{9\mu_l} = \left(1 + \frac{1}{\mathrm{e}}\right)\ln\left(\frac{2\,\delta}{3\mathrm{h_a}}\right) \Rightarrow \begin{Bmatrix} > 1\ \text{rebound} \\ < 1\ \text{adhesion} \end{Bmatrix} \tag{5.3}$$

where v_o is the velocity of particle collision, m and R are the particle mass and radius respectively, h_a is the height of the surface asperity, μ_l and δ are the viscosity and thickness of the liquid layer respectively, and e is the coefficient of restitution.

Simons et al. (1993) and Fairbrother (1999) considered the capillary static forces to derive a simple model capable of predicting the rupture energy of pendular liquid bridges, with only knowledge of the liquid volume employed to generate the liquid bridge itself:

$$\mathrm{W}^* = \mathrm{k}\,\mathrm{V}_\mathrm{b}^{*\,0.5} \tag{5.4}$$

where W^* is the dimensionless rupture energy ($W^* = W/\gamma\, R^2$, with γ the liquid surface tension), V^* is the dimensionless bridge volume ($V^* = V_b/R^3$) and k is a constant equal to 1.8. Their model predicts that the higher the gap value, the smaller the cohesive force between the particles.

Recently, Landi et al. (2011, 2012) validated the theories above in their investigation of the role of the interparticle forces on the flow behaviour of non-porous glass powders conditioned in a fluidized bed in controlled humid air at relative humidities between 13 and 98 %. Using the assumption that for non-porous materials, capillary condensation is the main phenomenon which is responsible for the formation of liquid bridges, they developed a model from shear experiments to predict the flow behaviour of the glass materials investigated and found that the tensile strength between the particles is a function of the cohesive force which, in turn, is a function of the bridge gap and of the asperity radius. In agreement with Simons et al. (1993) and Fairbrother (1999), the strength of the interparticle forces due to the capillary condensation between the asperities depended on the value of the capillary bridge gap.

5.2.5 *Solid Bridges: Sintering*

Temperature can have a considerable effect on particle adhesion if the contact between particles takes place at temperatures sufficiently high to cause softening of the particle surface and formation of interparticle bonds. The temperature at which softening occurs is called minimum sintering temperature, T_s. This is often lower than the fusion temperature of the bulk of the material.

The sintering process is characterised by the migration of particle material towards the bond zone. This can occur according to four different mechanisms, as described by Siegell (1984): surface diffusion, volume diffusion, viscous flow, vaporization. More than one mechanism can occur simultaneously depending on the material and on the conditions under which it is sintering. Transport of material by diffusion and viscous flow are considered the most important in defluidization phenomena.

Sintering by diffusion involves the movement of individual atoms from high to low density regions and consequently migration of lattice vacancies from regions of high to low vacancy concentration. Diffusion can occur both at the surface, surface diffusion, and through the bulk of the material, volume diffusion. Sintering by surface diffusion usually happens in the early stages of all sintering processes, and is the cause of the initial adhesion between particles, which leads to the formation of agglomerates. Surface diffusion is followed by volume diffusion which causes the densification of the material. Sintering by diffusion is typical of crystalline and metallic materials.

The mass transfer mechanism in sintering by viscous flow is described, on a microscopic level, as due to the movement of entire planes of lattice, as opposed to the movement of single atoms which occurs in the diffusion mechanism. Thus, the rate of growth of the bond area is higher if sintering is by viscous flow, and the agglomerates which are formed are much more strongly bonded than those caused by a diffusion mechanism. This mechanism is the most important in defluidization because it is the most rapid.

When two particles of a fluidized bed come in contact with each other at high temperature they will tend to form a bond. Defluidization of a fluidized bed will take place when the bonds caused by sintering cannot be broken apart by the kinetic motion of the particles in the bed. Strong agglomerates, difficult to break, are caused by densification of the bond zone, which is not only a function of the temperature but also of how long the particles remain bonded. The strength of the agglomerates which form during defluidization depends also on the sintering mechanism. Siegell (1984) observed that friable agglomerates are formed during sintering by diffusion, and that this mechanism does not alter the original shape of the particles.

Compo et al. (1987) have used thermo-mechanical analysis (TMA) to quantify the sintering temperature; they correlated the dimensionless excess velocity $(u - u_{mfs})/u_{mfs}$ with the dimensionless excess temperature $(T - T_s)/T_s$, where u_{mfs} is the minimum fluidization velocity at the minimum sintering temperature calculated from the value at ambient conditions using the Ergun equation. A theoretical model had been developed earlier by Tardos et al. (1985) to predict the limiting gas velocity U_s which is necessary to break the largest agglomerate in the bed and thereby to keep a bed of sticky particles continuously fluidized at temperatures above the minimum sintering temperature. The theoretical model was based on a force/stress balance on an agglomerate, cylindrical in shape and non-freely buoyant, which was assumed to occupy the entire cross-section area of the bed. The magnitude of the forces acting on the agglomerate, mainly due to the passage of bubbles, was estimated as a function of the excess fluidizing gas velocity, $u - u_{mf}$. The forces were then related to the pressure, q, acting on the agglomerate. Failure of the structure was predicted to occur when the pressure exceeded the maximum value q_{max} defined as:

$$q_{max} = \sigma_y \left(\frac{2h}{d_{ag}}\right)^2 A_1 \tag{5.5}$$

where σ_y is the yield strength of the agglomerate, d_{ag} and h are the diameter and height of the agglomerate and A_1 is a coefficient approximately equal to 2.

An alternative approach was proposed by Ennis et al. (1991). Their model is based on the concept that, when particles collide, kinetic energy is dissipated via viscous losses in the fluid in the contact zone. At low collision velocities all energy is dissipated and the particles adhere. Above a certain critical velocity insufficient energy is dissipated in the fluid and the particles rebound.

Seville et al. (1998) described the phenomenon of defluidization caused by visco-plastic sintering. In a simple model of a fluidized bed, the particles are considered to remain in quiescent zones with relatively little movement until they are disturbed by the passage of bubbles. If the residence time in the quiescent zones is sufficiently long for the sinter necks to reach a critical size such that the agglomerates cannot be broken by the passage of the bubbles, then defluidization will start occurring. Seville et al. (1998) modelled the sintering phenomenon on the basis of a comparison of the characteristic residence time in which the particle motion is relatively small, t_{bb}, and the characteristic time necessary for the growth of sinter necks, t_s. The latter will change with temperature. In this approach, the time spent in the quiescent zone was assumed to be a function of the excess fluidizing velocity:

$$t_{bb} = \frac{K_1}{(u_{mfs} - u_{mf})} \tag{5.6}$$

where K_1 is a constant which equals $2d_b/3$ when t_{bb} is considered the average time between the passage of bubbles, and d_b is the bubble diameter.

The critical time for sintering, t_s, sufficient to form an agglomerate which cannot be broken by the bubble was expressed as:

$$t_s = \left(\frac{x}{r}\right)^2 \frac{\eta}{k_1} \tag{5.7}$$

where x is the neck radius at time t, r is the radius of the particle, η is the surface viscosity and k_1 is a factor dependent on both materials' properties and environmental conditions. Thus, Seville et al. (1998) obtained a quantitative relationship between the velocity required to keep the bed fluidized and bed temperature in terms of the surface viscosity of the particles by equating (5.6) and (5.7).

In order to predict the defluidization behaviour of a fluidized bed it is necessary to determine the initial sintering temperature of the particles. Like Compo et al. (1987), Lettieri (1999) measured the minimum sintering temperature of a range of materials using dilatometry analysis. Lettieri (1999) used thermomechanical analysis (TMA) to determine the expansion/contraction mechanisms taking place when

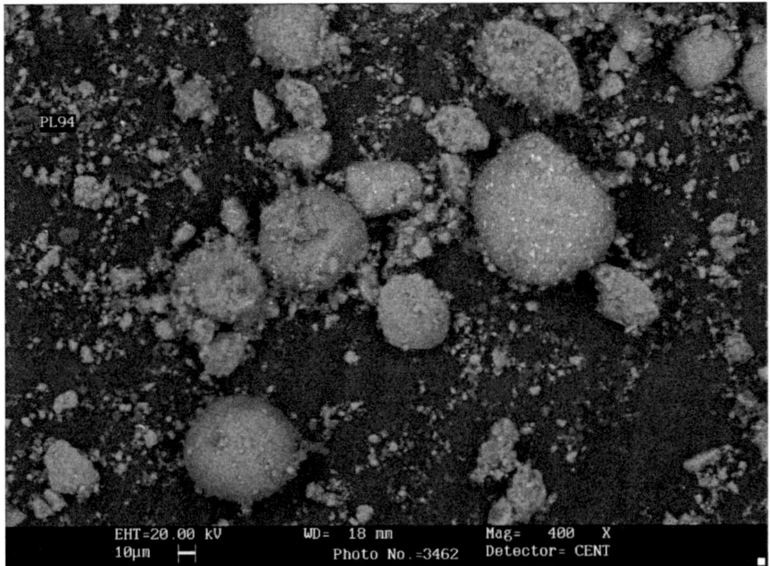

Fig. 5.1 SEM of an E-cat prior to high temperature fluidization (Lettieri 1999)

heating up samples of different industrial catalysts in order to relate changes in the materials' properties to their fluidization behaviour with increasing temperature. Figure 5.1 represents an SEM analysis of a sample of an equilibrium-catalyst (E-cat) prior to high temperature fluidization, showing a large number of fines stuck onto the surface of the larger particles. Figure 5.2, shows the effect of sintering after the powder was fluidized at high temperature, where strong bonds formed between the fine particles giving place to large agglomerates. The samples analyzed contained largely Si, Al, O and some C. Carbon was present on both the fines and the larger particles, but particles with lower levels of adhered fines appeared to contain less carbon.

The results of TMA carried out using a dilatometer are shown in Fig. 5.3, in which changes in the equilibrium catalyst dimension with increasing temperature are reported. Figure 5.3 shows an initial expansion up to 134 °C after which a sharp decrease in size occurs up to about 200 °C. The thermogram also shows a second very sharp shrinkage between 414 and 429 °C, after which the particle size remained constant until 900 °C, when a small amount of shrinkage takes place and which continued until the end of the experiment. A quantitative analysis of the TMA results showed a relative increase in size of 1.2 % occurred while heating the sample up to 132 °C. This was followed by a decrease in size of about 10 % between 132 and 250 °C. An even more important dimensional change occurred during the small temperature range between 414 and 429 °C where a relative size decrease of 11 % was quantified. Each shrinkage corresponds to sintering taking

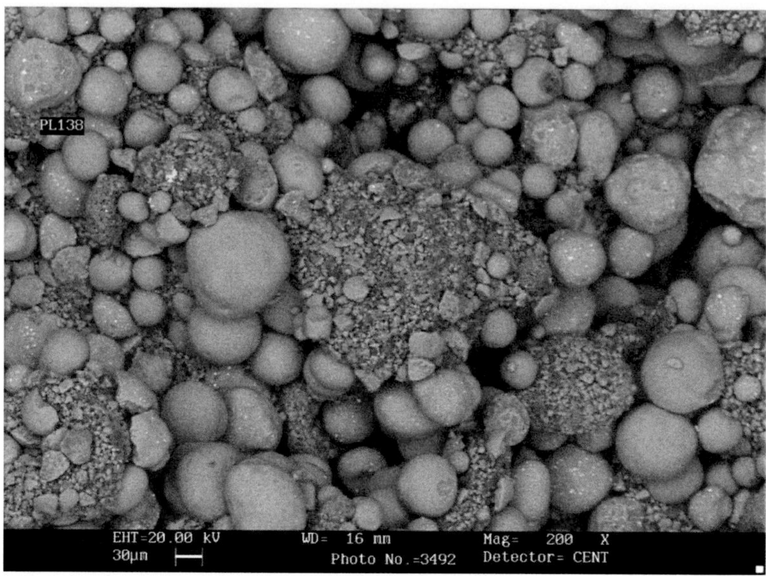

Fig. 5.2 SEM of an E-cat after high temperature fluidization (Lettieri 1999)

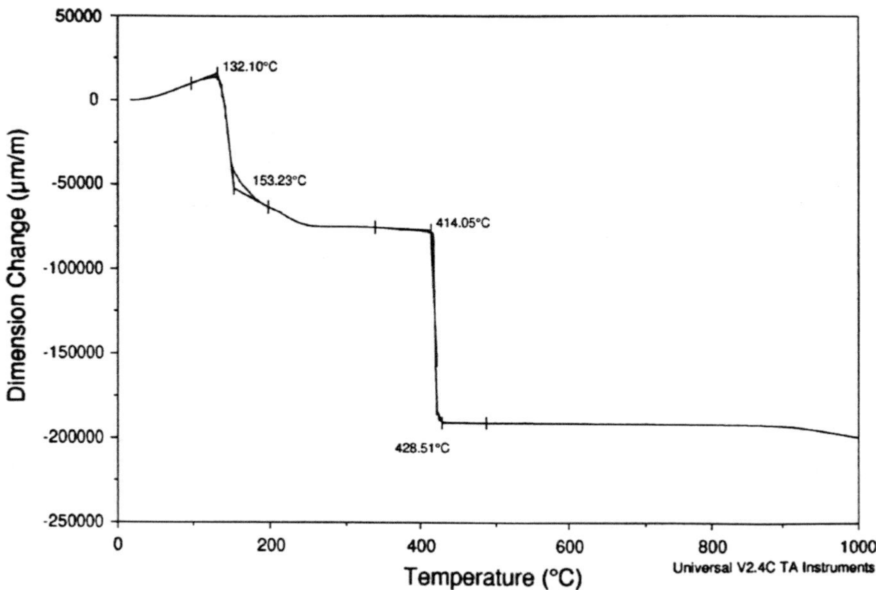

Fig. 5.3 TMA of an E-cat

place, in agreement with complete defluidization being observed between 150 and 200 °C (Lettieri 1999).

5.3 Effect of Temperature on Fluidization

The process conditions influence the operation of fluid-particle systems because they affect gas density and viscosity. Increasing temperature causes gas density to decrease and gas viscosity to increase. As mentioned before, most predictions of the fluidization behaviour at high temperatures have been based solely on considering such changes in the gas properties. However, this approach is valid under the condition that only hydrodynamic forces control the fluidization behaviour. Temperature can have a considerable effect on particle adhesion, enhancing the role of the IPFs on the fluidization quality, if the system is operated at temperatures close to the minimum sintering temperature of the particles, as discussed in Sect. 5.2.5. The effect of temperature on a fluidized bed is also strongly dependent on particle size, which in turn defines the type of particle-particle and fluid-particle interaction, thus determining the stronger or weaker role of the IPFs.

5.3.1 Effect of Temperature on Minimum Fluidization Conditions

The correlation most widely used to predict u_{mf} at ambient temperature is the Ergun equation, which is an expression for the pressure drop through a settled bed of solids:

$$(\rho_s - \rho_f)\, g\, (1 - \varepsilon_{mf})\, L = 150\, \frac{\mu\, L\, u_{mf}(1 - \varepsilon_{mf})^2}{(\phi\, d_p)^2\, \varepsilon_{mf}^3} + 1.75\, \frac{\rho_g\, L\, u_{mf}^2\, (1 - \varepsilon_{mf})}{\phi\, d_p\, \varepsilon_{mf}^3} \tag{5.8}$$

In order to solve Eq. 5.8 the value of the bed voidage at minimum fluidization, ε_{mf}, and the sphericity of the particles, ϕ, need to be known a priori.

Wen and Yu (1966) showed that the voidage and shape factor functions in both the viscous and the inertial term of Eq. (5.8) can be approximated as:

$$\frac{1 - \varepsilon_{mf}}{\phi^2\, \varepsilon_{mf}^3} \approx 11; \qquad \frac{1}{\phi\, \varepsilon_{mf}^3} \approx 14 \tag{5.9}$$

Combining Eqs. (5.8) and (5.9) Wen and Yu expressed the Ergun equation as follows:

$$\mathrm{Ga} = 1650\mathrm{Re}_{\mathrm{mf}} + 24.5\mathrm{Re}_{\mathrm{mf}}^2 \tag{5.10}$$

From the viscous term of Eq. (5.10) u_{mf} for small spherical particles (below about 100 μm) is given by:

$$u_{mf} = \frac{d_p^2(\rho_p - \rho_f)\,g}{1650\mu} \tag{5.11}$$

For larger particles, it becomes:

$$u_{mf}^2 = \frac{d_p(\rho_p - \rho_f)\,g}{24.5\,\rho_f} \tag{5.12}$$

Referring to any particle system belonging to any of the Geldart Groups, the qualitative effect of temperature on u_{mf} can be predicted from considerations on the gas density and viscosity terms in the Wen and Yu equation. For small particles, Eq. (5.11) shows that u_{mf} varies with $1/\mu$. Therefore, u_{mf} should decrease as temperature increases, when the viscous effects are dominant. Equation (5.12) predicts that u_{mf} will vary with $(1/\rho_f)^{0.5}$, thus u_{mf} should increase with temperature for large particles, when turbulent effects dominate. However, predictions with Eqs. (5.11) and (5.12) do not take into account possible changes in the voidage which may occur with increasing temperature. Various other correlations can be found in the literature to predict u_{mf} at high temperature, see Table 5.2.

Experimental verification of the temperature effect on u_{mf} has been reported by several authors. As predicted by the Wen and Yu equation, Botterill et al. (1982) observed a decrease of u_{mf} with increasing temperature for Group B materials, because of the consequent increase in gas viscosity, whereas for the large Group D powders they observed an increase in u_{mf}, because of the decrease of gas density, with the voidage at minimum fluidization being independent of temperature. The latter has been the subject of further experimental observations, Lucas et al. (1986), Raso et al. (1992), Formisani et al. (1998), and Lettieri et al. (2001a, b) all observed changes in the voidage at minimum fluidization with increasing temperature. Controversy is however reported on the phenomena which determine such changes. Lucas et al. (1986) explained changes in ε_{mf} with temperature on a hydrodynamic basis, suggesting a change in the flow pattern inside the bed. Contrary to this, Raso et al. (1992), Yamazaki et al. (1995), Formisani et al. (1998) and Lettieri et al. (2001a, b) related such changes to a variation of interparticle forces with increasing temperature. In particular, Formisani et al. (1998) investigated various Group A, B and D powders and observed a linear increase of the voidage of the fixed bed with temperature and a corresponding linear increase of ε_{mf} (Fig. 5.4) with a close similarity between the slope of the fixed bed voidage and the voidage at minimum fluidization. In line with the theory previously advanced by Rietema, they attributed the increase of the fixed bed voidage by assuming that the interparticle forces between cohering particles give rise to a powder structure with a certain mechanical

Table 5.2 Selected equations for the calculation of the minimum fluidization velocity, u_{mf}

Authors	Equation
Ergun (1952)	$150\frac{\mu_g u_{mf}}{(\phi d_p)^2}\frac{(1-\varepsilon_{mf})}{\varepsilon_{mf}^3}+1.75\frac{\rho_g u_{mf}^2}{\phi d_p}\frac{1}{\varepsilon_{mf}^3}=g(\rho_p-\rho_g)$
Carman (1937)	$u_{mf}=\frac{(\phi d_p)^2}{180}\frac{(\rho_p-\rho_g)}{\mu_g}g\left(\frac{\varepsilon_{mf}^3}{1-\varepsilon_{mf}}\right)$
Miller and Logwinuk (1951)	$u_{mf}=\frac{1.25\times10^{-3}d_p^2(\rho_p-\rho_g)^{0.9}\rho_g^{0.1}g}{\mu_g}$
Leva et al. (1956)	$u_{mf}=\frac{7.39d_p^{1.82}(\rho_p-\rho_g)^{0.94}}{\rho_g^{0.06}}$
Goroshko et al. (1958)	$u_{mf}=\frac{\mu_g}{\rho_g d_p}\left(\frac{Ar}{1400+5.2\sqrt{Ar}}\right)$
Leva (1959)	$u_{mf}=\frac{8.1\times10^{-3}d_p^2(\rho_p-\rho_g)g}{\mu_g}$
Broadhurst and Becker (1975)	$u_{mf}=\frac{\mu_g}{\rho_g d_p}\left(\frac{Ar}{2.42\times10^5Ar^{0.85}\left(\frac{\rho_p}{\rho_g}\right)^{0.13}+37.7}\right)^{0.5}$
Riba et al. (1978)	$u_{mf}=\frac{\mu_g}{\rho_g d_p}\left(1.54\times10^{-2}\left(\frac{d_p^3\rho_g^2 g}{\mu_g^2}\right)^{0.66}\left(\frac{\rho_p-\rho_g}{\rho_g}\right)^{0.7}\right)$
Doichev and Akhmakov (1979)	$u_{mf}=\frac{\mu_g}{\rho_g d_p}(1.08\times10^{-3}Ar^{0.947})$
Wu and Baeyens (1991)	$u_{mf}=\frac{\mu_g}{\rho_g d_p}\left(7.33\times10^{-5}\times10^{\sqrt{8.24\log_{10}Ar-8.81}}\right)$

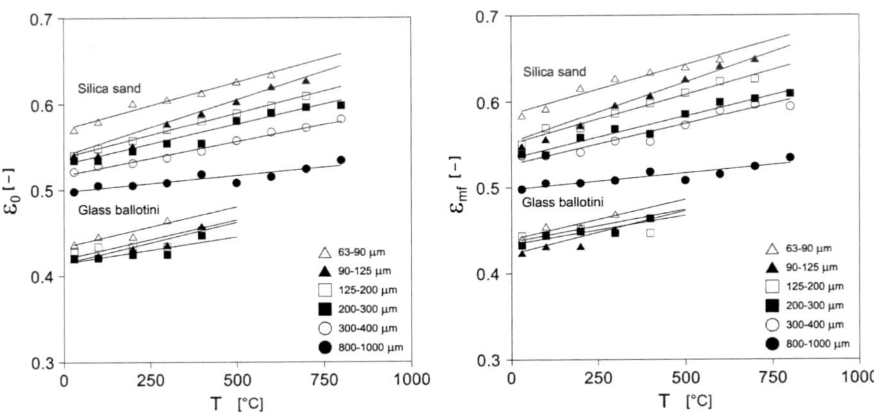

Fig. 5.4 Effect of temperature on the fixed bed voidage and minimum fluidization velocity for glass ballotini and silica sand particles of different size, Formisani et al. (1998)

strength in the packed bed and in turn in the bed at incipient fluidization and in the expanded state of homogeneous fluidization.

Lettieri (1999) reported on the effect of temperature on the minimum fluidization velocity of fresh industrial catalysts and catalysts doped with potassium acetate. The variation of minimum fluidization velocity with temperature was found to be

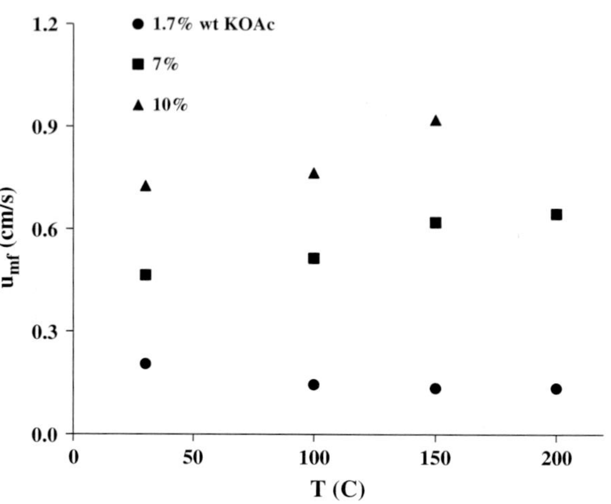

Fig. 5.5 Effect of temperature on the minimum fluidization velocity for an industrial catalyst doped with different levels of potassium acetate

sensitive to whether the HDFs or IPFs dominated the fluidization behaviour. For all fresh catalysts, values of the minimum fluidization velocity were predicted by the viscous dominated term of the Ergun equation, once appropriate values for the sphericity factor and ε_{mf} were used. However, u_{mf} values obtained for the doped catalysts increased as a function of temperature, and the values for u_{mf} were found to deviate from the predictions with the Ergun equation due to a stronger role of the IPFs. Figure 5.5 shows the minimum fluidization velocity of the three doped silica catalysts as a function of temperature. u_{mf} decreased slightly with increasing temperature, when the catalyst was doped with only 1.7 %wt of potassium acetate. Values of u_{mf} obtained for the sample with 7 %wt remained fairly constant up to 100 °C, then increased slightly between 100 and 200 °C. A greater increase of u_{mf} as temperature increased was found for the sample doped with 10 %wt of potassium acetate.

More recently, several authors have investigated also the combined effects of temperature and particle size and particle size distribution (PSD) on minimum fluidization velocity, Lin et al. (2002), Bruni et al. (2006), Subramani et al. (2007), Hartman et al. (2007), Goo et al. (2010), Chen et al. (2010) and Jiliang et al. (2013). General observations demonstrated that that operating temperature and particle size distribution can influence the minimum fluidization velocity simultaneously, making variations of u_{mf} non-monotonic with temperature. Several correlations have been derived for the prediction of the minimum fluidization conditions at high temperature, these are however case specific. The debate of the phenomena causing changes in behavior with increasing temperature remains controversial with still much disagreement on the role of the hydrodynamic and interparticle forces.

5.3.2 Effect of Temperature on Fluid-Bed Expansion and Richardson-Zaki Relationship

As mentioned earlier, Group A particles are those which exhibit a region of uniform expansion for gas velocities above minimum fluidization. The non-bubbling expansion is characterized by the Richardson-Zaki (1954) equation (Chap. 1, Eq. 1.1), which was first used to correlate the homogeneous expansion of liquid fluidized beds. Bed expansion experiments in gas-solid fluidized beds were conducted by Godard and Richardson (1968) on various materials, characterized by a very narrow size distribution, and fluidized with air at pressures between 1 and 14 atm. They found that the relationship between the fluidizing velocity u and bed voidage ε could be expressed in the form of Eq. (1.1):

$$U/U_t = \varepsilon^n \qquad (1.1)$$

The applicability of Eq. (1.1) to describe the bed expansion of Group A powders may suggest that the expansion mechanism of gas fluidized beds and liquid fluidized beds is similar. However, the validity of this comparison has not always been accepted. Massimilla et al. (1972) and Donsi' and Massimilla (1973) made some observations of the bubble free expansion of gas fluidized beds of fine particles and described the mechanism of bed expansion as due to nucleation and growth of cavities whose size ranges in the order of few particle diameters. At the same time, they also postulated that particles surrounding the cavities maintain the surface contacts, which is essential for the stability of the structure. They stated that the cavity growth mechanism of bed expansion probably occurs because of a broad distribution of interparticle forces.

This was evidenced by the different values of n found when comparing liquid fluidized beds and gas-solid systems. If for liquid systems, values of n were found to be equal to 4.8 in the viscous flow regime and 2.4 in the inertial regime, the values of the index n for powders were found to be higher than those predicted for uniform spheres fluidized by a liquid.

Various authors found that experimental values of n (indicated as n^*) extrapolated from expansion profiles are greater than those predicted by the Richardson-Zaki correlations. Some of the data reported in the literature are shown in Fig. 5.6.

Godard and Richardson (1968) found values of n^* between 4.7 and 8.9 for various materials fluidized with air at ambient conditions, the highest values were obtained for some phenolic resins. Massimilla et al. (1972) found values between 5.4 and 7, where the highest values were obtained for the finer and non-sieved materials. Geldart and Wong (1984, 1985) fluidized a wide range of powders at ambient conditions using various gases, such as air, argon, nitrogen and Arcton-12, and found values of n between 4 and 60, where the discrepancy becomes increasingly larger for those materials which showed higher degrees of cohesiveness. Similar results were also reported by Avidan and Yerushalmi (1982). Foscolo et al. (1987) reported values of n^* close to the predicted ones for particles having a

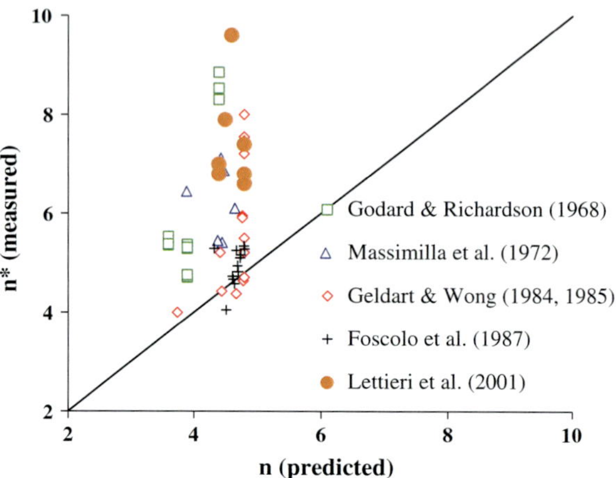

Fig. 5.6 Comparison between experimental n^* values obtained by various authors, at ambient conditions using different gases, with calculated n values using Richardson-Zaki correlations, Eq. (1.1)

very narrow particle size distribution and fluidized with air, argon and CO_2. This was in agreement with the findings of Lettieri et al. (2001a, b) who investigated the effect of temperature on the expansion profiles of four FCC powders. Experimental values of n^* and u_t^* were determined from the expansion profiles plotted in the Richardson-Zaki form and found to be greater than those predicted, with values being within the range 6.4–9.6.

The discrepancy concerning u_t^* might be partly explained by the large extrapolation in the data that must be employed, i.e. from $\varepsilon = 0.6$ to $\varepsilon = 1$. However, Avidan and Yerushalmi (1982) stressed the great influence that the particle size distribution may have on the values obtained for u_t^*. They found lower u_t^* values for those catalysts characterized by a higher content of fines. This was in agreement with the results by Lettieri et al. (2000), where values obtained for the FCC3, which contained about 25 % of fines, were lower than those obtained for the other FCC catalysts containing respectively 5 and 16 % of fines.

Furthermore, it is important to mention that it is difficult to know which mean particle diameter to use when calculating u_t for powders with a wide particle size distribution. In fluidization, the surface-volume ratio, d_{SV}, is generally accepted as the most appropriate estimate of the mean particle diameter. However, if other possible geometrical diameters are considered, such as the surface average and volume average, d_S and d_V respectively:

$$d_S = \sqrt{\sum x_i d_i^2} \qquad d_V = \sqrt[3]{\sum x_i d_i^3} \tag{5.13}$$

where x_i is the mass fraction of particles in each size range given by the sieve aperture d_i, then the mean particle diameters for the three samples of fresh FCC catalysts calculated as d_{SV}, d_S and d_V are reported in Table 5.3.

Lettieri (1999) calculated the particle terminal fall velocity corresponding to the diameters in Table 5.3 for three FCC catalysts, and compared such values against

Table 5.3 Mean particle diameters

	d_{SV} (μm)	d_S (μm)	d_V (μm)
FCC 1	71	91	102
FCC 2	57	104	124
FCC 3	49	80	91

those obtained experimentally, and found that the values of u_t^* extrapolated at temperatures above 100 °C were of the same order of magnitude as u_{tS} and u_{tV}. For all FCC catalysts, u_t^* values obtained between 20 and 100 °C corresponded to a mean particle diameter much greater than either d_S or d_V.

Also Valverde et al. (2001) investigated the role of the interparticle forces on the homogenous fluidization and settling of fine powders. They proposed an extension of the Richardson-Zaki empirical correlation to predict the effect of the interparticle forces on the settling of fine powders in the presence of aggregates. Valverde et al. (2003) extended the previous study investigating the transition between the solid-like, fluid-like, and bubbling fluidization of gas-fluidized fine powders. Using optical probe measurements, they showed that the transition between the solid-like and the fluid-like regimes takes place along an interval of gas velocities in which transient active regions alternate with transient solid networks, making the prediction of the transition between the different regimes a complex task. Castellanos (2005) later studied the onset of fluidization of fine and ultrafine powders and attributed to the presence of clusters the observation of a highly expanded state of uniform fluid-like fluidization. Valverde and Castellanos (2008) combined the observations reported above proposing an extension of the Geldart classification of powders to predict the behavior of gas-fluidized cohesive particles taking into account interparticle forces. In the new diagram proposed by Valverde and Castellanos, the boundaries between the different types of fluidization are not defined solely by hydrodynamic and physical parameters such as fluid viscosity and particle density but are also a function of the fractal dimension of the agglomerates and the powder's compaction history.

5.3.3 *Effect of Temperature on the Stability of Group A Powders*

The transition from homogenous to bubbling fluidization can be predicted using the stability criterion developed by Foscolo and Gibilaro (1984), previously introduced in Chap. 1. We summarize here the fundamental assumptions of the model:

- hydrodynamic forces, i.e. gravitational force, buoyancy and drag force, control the stability of Group A powders, at both ambient and high temperatures.
- the buoyancy force, W_b, exerted on a particle is defined as a function of the density of the suspension, rather than of the fluid alone.
- the Richardson-Zaki equation is used to describe the relation between the velocity and voidage, with values of n = 4.8 for the viscous regime.
- by applying Richardson-Zaki equation, the pressure drop is expressed as $\Delta P \propto \varepsilon^{-4.8}$, and the drag force is given by $F_d \propto \varepsilon\Delta P \propto \varepsilon^{-3.8}$.
- Wallis' stability theory is applied to determine the transition between particulate and bubbling regime, determined as ε_{mb}.

The expression of the criterion was given in Chap. 1, Sect. 1.2.5, we write it again for convenience:

$$\left[\frac{g\,d_p(\rho_p-\rho_f)}{u_t^2\,\rho_p}\right]^{0.5} - 0.56\,n\,(1-\varepsilon_{mb})^{0.5}\,\varepsilon_{mb}^{n-1} = \left\{\frac{\text{positive,stable}}{\text{negative,unstable}}\right\} \tag{5.14}$$

The Foscolo-Gibilaro model was developed for systems of spherical mono-sized particles, for which the values of n and u_t can be adequately predicted by the Richardson-Zaki correlations and the Stokes law, based on the surface-volume diameter.

Lettieri et al. (2001a, b) validated Eq. (5.14) for different FCC catalysts fluidized at high temperature and found that prediction of ε_{mb} didn't match with the experimental evidence, as the stability criterion predicted a much greater increase in the voidage at minimum bubbling than observed. In the original Foscolo-Gibilaro stability criterion, the constitutive equation for the interaction force on a single particle is expressed as the sum of the contribution given by the buoyancy force and by the drag force. The latter determines the homogeneous expansion of the bed, through its relation with n and u_t. Thus, given the discrepancy between the experimental n^* and u_t^* values shown in Figs. 5.6 and 5.7, Lettieri et al. (2001a, b) proposed a generalization of the stability criterion by re-formulating the drag force by imposing a general value n for the Richardson-Zaki index in the expression of the drag. Hence they obtain a generalized stability criterion that can be expressed as follows:

$$\left[\frac{2\,g\,d_p\,(\rho_p-\rho_f)}{3\,u_t^2\,\rho_p}\right]^{0.5} - n^{0.5}\,(1-\varepsilon_{mb})^{0.5}\varepsilon_{mb}^{n-1} = \left\{\frac{\text{positive,stable}}{\text{negative,unstable}}\right\} \tag{5.15}$$

The generalized model, Eq. (5.15), allows to use values of n and u_t which can be different from those originally proposed. A comparison between predictions obtained from the original and generalized Foscolo-Gibilaro model for three fresh FCC, from 20 to 650 °C are reported in Fig. 5.7. These results demonstrate clearly the importance of using a relation between the drag force and the expansion parameters which correctly describes the characteristics of the homogeneous fluid-bed system. When n^* and u_t^* values, that characterize the bed expansion, are

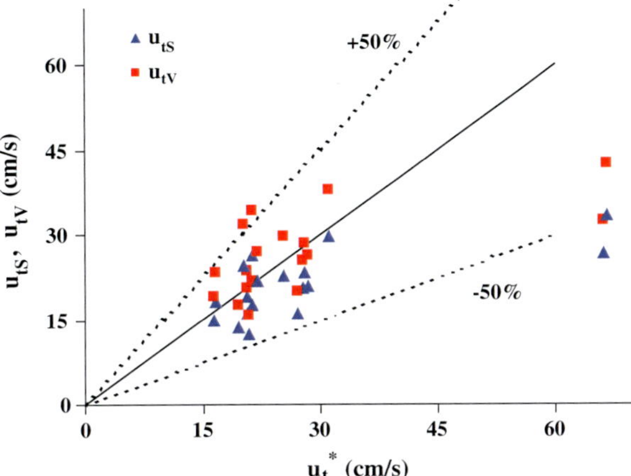

Fig. 5.7 Experimental u_t^* values *versus* calculated values of u_{tV} and u_{tS} for three fresh FCC catalysts, from 100 to 650 °C (Lettieri 1999)

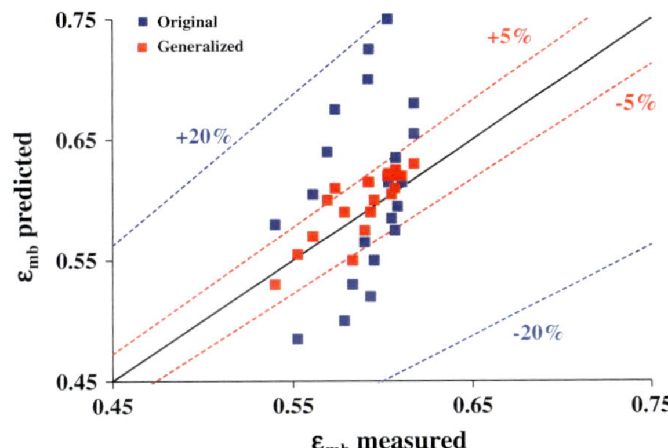

Fig. 5.8 Comparison between predictions obtained from the original and generalized Foscolo-Gibilaro model for three fresh FCC, from 20 to 650 °C (Lettieri 1999)

introduced in the Foscolo-Gibilaro model, this is capable of predicting the transition between the particulate and bubbling regime with a smaller margin of error (i.e. ±5 %), than the original model (i.e. ±20 %) (Fig. 5.8).

5.3.4 *Effect of Temperature on the Non-bubbling Ratio*

The non-bubbling ratio, u_{mb}/u_{mf}, is reported to be one of the key parameters which characterizes the fluidization of fine materials. It is used as a measure of how fluidized beds expand or contract; the larger the u_{mb}/u_{mf} ratio the smoother the fluidization quality, and the better the aeratability of the materials. Abrahamsen and Geldart (1980) related the hydrodynamic properties of fluidized beds to the non-bubbling ratio. They performed measurements of the u_{mb}/u_{mf} ratio over a wide range of materials at ambient conditions using different gases. They proposed a correlation to predict the non-bubbling ratio from the properties of the powder, i.e. particle diameter and fines content, and from the properties of the fluidizing gas, i.e. density and viscosity.

$$\frac{u_{mb}}{u_{mf}} = \frac{2300\,\rho_g^{0.126}\,\mu^{0.523}\,\exp\,(0.716\,F_{45})}{g^{0.934}\,(\rho_p - \rho_g)^{0.934}\,d_p^{0.8}} \tag{5.16}$$

According to the Abrahamsen and Geldart (1980) correlation, u_{mb}/u_{mf} varies as a function of the properties of the powder when the gas properties are not changed. For example, an increase in the fines contents due to attrition, will cause u_{mb}/u_{mf} to increase and the fluidization quality to improve. However, if the fines content increases too much, the fluidization behaviour could eventually shift from Group A type to cohesive Group C. Conversely loss of fines, which may occur through mal-functioning of a cyclone, reduces u_{mb}/u_{mf}. When u_{mb}/u_{mf} becomes close to 1, the flow behaviour can shift from a Group A into a Group B type fluidization.

The Abrahamsen and Geldart (1980) correlation predicts the effect of operating conditions on fluidization quality through changes in the gas density and viscosity terms ($u_{mb}/u_{mf} \sim \mu^{0.523} \cdot \rho^{0.126}$), assuming that only HDFs are present. As temperature increases, changes in the viscosity term dominate, and the correlation predicts that u_{mb}/u_{mf} should increase, thus improving the fluidization quality.

Newton et al. (1996) reported on the effect of temperature on the u_{mb}/u_{mf} ratio of some fresh FCC catalysts, which were fluidized in a 100 mm i.d. vessel from ambient conditions up to 500 °C. They observed a decrease of the u_{mb}/u_{mf} ratio with increasing temperature for all FCC catalysts. The experimental values were compared with the predictions given by the Abrahamsen and Geldart (1980) correlation. Experimental trends were found to be opposite to the predicted ones. Furthermore, Newton et al. (1996) observed that differences existing between the catalysts at ambient conditions, in terms of the u_{mb}/u_{mf} ratio, disappeared at high temperature, altering the ranking order of the powders.

Xie and Geldart (1995) proposed Eq. (5.17), a slightly modified version of Eq. (5.16), which was developed on the basis of tests carried out also at high temperature:

$$\frac{u_{mb}}{u_{mf}} = \frac{333\,\rho_g^{0.19}\,\mu_g^{0.37}\exp\,(0.716\,F_{45})}{g^{0.934}\,(\rho_p - \rho_g)^{0.934}\,d_p^{0.8}} \tag{5.17}$$

They also proposed a further version, Eq. (5.18) which was obtained by considering that u_{mb} changes with temperature are proportional to $\rho_g^{0.13}/\mu^{0.5}$, as shown in Eq. (5.15):

$$u_{mb} = 0.3\exp\,(0.716\,F_{45})\frac{d_p\,\rho_g^{0.13}}{\mu^{0.5}} \tag{5.18}$$

and that u_{mf} is inversely proportional to μ, see Ergun (Table 5.2). Then, by taking gas density as inversely proportional to absolute temperature, and gas viscosity as proportional to the square root of absolute temperature, changes in u_{mb}/u_{mf} were expressed as proportional to $T^{0.12}$, and the following correlation was proposed:

$$\left(\frac{u_{mb}}{u_{mf}}\right)_T = \left(\frac{u_{mb}}{u_{mf}}\right)_{297} \left(\frac{T}{297}\right)^{0.12} \tag{5.19}$$

Note that an experimental value of the non-bubbling ratio at ambient temperature is required in this equation.

As temperature increases the gas density decreases whilst gas viscosity increases. It is therefore predicted that both u_{mf} and u_{mb} should decrease, with u_{mf} decreasing faster than u_{mb} with increasing temperature. Thus, Eq. (5.19) predicts that the u_{mb}/u_{mf} ratio should increase. Lettieri (1999) tested all three equations: Eqs. (5.16), (5.17) and (5.19) for three FCC catalysts, and found that Eq. (5.16) predicted a much greater increase of the non-bubbling ratio than the one found experimentally. Thus, extrapolating the effect of temperature on the fluidization of these catalysts from Eq. (5.16), may lead to a misleading prediction of the non-bubbling ratio at high temperature. On increasing temperature, Eq. (5.19) gave a better prediction of u_{mb}/u_{mf} values for two of the FCC catalysts with increasing temperatures. On the whole, all of the equations gave predictions with a scatter of ±30 %, as shown in Fig. 5.9.

More recently, Girimonte and Formisani (2009) investigated the influence of operating temperature on the transition to the bubbling regime for some FCC, silica and corundum sands, at temperatures ranging from 30 to 500 °C. They determined the minimum bubbling velocity using four different methods and showed that depending on the method adopted, different results can be obtained for u_{mb} with increasing temperature. The first method relied on the classical direct observation of the velocity at which the first bubble erupted on the free surface of the bed. The second method was based on the measurement of the pressure drop across the

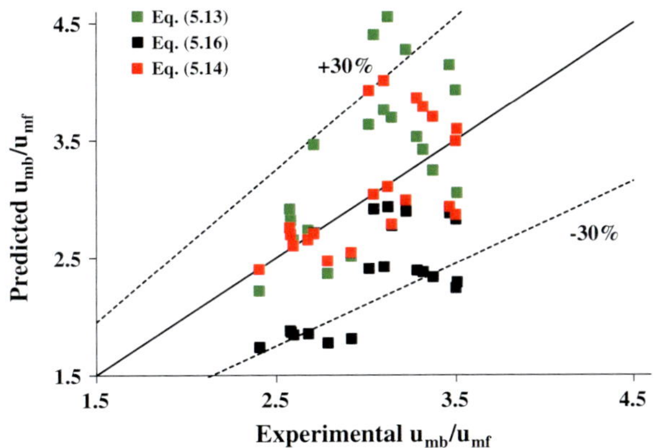

Fig. 5.9 Non-bubbling ratio, experimental *versus* predicted values for three FCC catalysts from 20 to 650 °C (Lettieri 1999)

whole bed, and u_{mb} was taken at the point where a shallow minimum of the Δp versus u curve occurs. The last two methods were derived from the analysis of the "fluidization map", namely the examination of the expansion behaviour of the bed over a range of fluidization velocities from the fixed bed state to the bubbling regime. Based on the experimental evidence, Girimonte and Formisani (2009) concluded that the visual observation of the bed and the method based on the detection of the pressure drop minimum were unreliable for correctly determining the starting point of bubbling. They concluded that the analysis of bed expansion as a function of the fluidization velocity is the only method allowing the reconstruction of the succession of phenomena through which a stable flow of bubbles across the solid mass ensues.

Girimonte and Formisani (2014) reported in further experiments on the effect of temperature on the fluidization of FCC particles. They used a combination of non-invasive optical technique for acquiring images of bubbles' eruption at the free surface and results from bed collapse tests to determine the transition to bubbling fluidization with increasing temperature. Their experiments showed that high temperature influences the quality of bubbles producing a smoother regime of bubbling, which they attributed to the thermal enhancement of IPFs that leads to higher porosity and lower interstitial flow in the emulsion phase.

In summary, high temperature clearly affects the stability of fluidized beds of Group A powders; well established theories and models fail to predict correctly the voidage at minimum bubbling with increasing temperature. Models corrected on the basis of experimental data are capable to reproduce correct trends; however a priori predictions of the fluid bed stability with increasing temperature are yet to be achieved. The challenge here still is in the ability to describe the forces that determine the transition from particulate to bubbling fluidization. Hence, some kind of quantification of the effects of the IPFs on fluidization is needed in order to advance the understanding of fluidization at high temperature.

5.4 Pressure

A number of fluidized-bed processes are operated at elevated pressures, gasification and polymerization being two examples. It is therefore important to know how fluidized beds behave under these conditions and how this behaviour differs from that observed at ambient pressure. Important to consider are the properties of beds over the full range of gas velocities from minimum fluidization through the bubbling regime to turbulent flow and velocities at which particle elutriation occurs. The effect of pressure on jet penetration length from immersed orifices and bed-to-surface heat transfer is also crucial for design and is discussed in what follows.

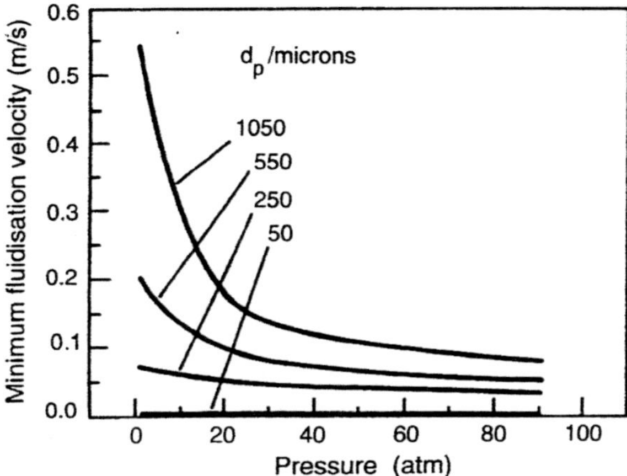

Fig. 5.10 Effect of pressure on U_{mf} based on Eq. 5.20 (Based on Rowe 1984)

5.4.1 Minimum Fluidization Velocity

The effect of pressure on U_{mf} may be estimated qualitatively by rearranging the Ergun equation as follows:

$$U_{mf} = \frac{\mu}{\rho_f d_p} 42.9(1 - \varepsilon_{mf}) \left\{ \left[1 + 3.0 \times 10^{-4} \frac{\varepsilon_{mf}^3}{(1 - \varepsilon_{mf})^2} Ar \right]^{1/2} - 1 \right\} \tag{5.20}$$

Rowe (1984) applied Eq. 5.20 to particles of density 1250 kg/m^3, a range of sizes and a value of ε_{mf} of 0.5 (Fig. 5.10).

It is clear that for particles with diameters less than 100 μm (Group A) pressure is predicted to have little effect, the reason being that gas flow around these small particles is laminar and the fluid-particle interaction force is dominated by gas viscosity which is largely independent of pressure in the range considered. With increasing particle size inertial forces become more important and at d_p > 500 μm (Group B) they begin to dominate over the viscous forces causing U_{mf} to decrease sharply with pressure up to about 20 bar and more gradually thereafter. King and Harrison (1982) also showed that U_{mf} is independent of pressure for laminar flow ($Re_{mf} < 0.5$) while for turbulent flow ($Re_{mf} > 500$) it is inversely proportional to the square root of gas density and hence pressure. Similar conclusions were reached by Olowson and Almstedt (1991) who measured U_{mf} for a range of particles in Groups B and D at pressures from 0.1 to 0.6 MPa and found a general decrease with increasing pressure.

5.4.2 Bubble Dynamics

A great deal of work has been carried out on the effect of pressure on bubbling beds much of which was summarised by Yates (1996) and subsequently (Yates 2003).

As in the case of minimum fluidization velocity the behaviour is a function of the type of bed material divided amongst the four Geldart groups.

5.4.2.1 Group A Materials

It is generally agreed that while U_{mf} is unaffected the region of bubble-free bed expansion between U_{mf} and U_{mb} increases with increasing pressure. In addition at the same values of volumetric gas flow rate bubbles in beds of Group A materials become smaller as pressure increases. There could be two reasons for this: (a) a greater proportion of gas flows through the emulsion phase as a result of an increase in emulsion-phase voidage; (b) the stability of bubbles decreases causing them to break up into smaller voids. The question of bubble stability has been considered since the early days of fluidization and two models have emerged from these studies. In the theory of Davidson and Harrison (1963) it was assumed that as the bubble rises the shear force exerted by the particles moving down relative to the bubble sets up a circulation of gas within the void with a velocity u_c which approximates to the bubble rise velocity u_b. When, through coalescence, the bubbles grow in size and their velocity increases a point is reached where u_c exceeds the terminal fall velocity, u_t, of the particles within the bubble and solids in the wake will be drawn up causing the bubble to break into smaller units with lower rise velocities. Bubbles would therefore be expected to reach a limiting size determined by u_t and beds of Group A particles should show "smoother" fluidization than beds of coarser, denser materials and since values of u_t decrease with increasing pressure (Haider and Levenspiel 1989) this behaviour should increase with pressure, an effect widely reported in the literature (Yates 1996).

An alternative theory proposes that bubble break-up is caused by a Taylor instability in the bubble roof allowing particles to rain down through the void and divide it in two (Clift et al. 1974). A factor determining the stability of the bubble roof is taken to be the apparent kinematic viscosity of the emulsion phase so that bubbles become more unstable as this viscosity decreases a change which would result from an increase in emulsion-phase voidage a trend already noted to occur with increasing pressure. An X-ray study by King and Harrison (1980) of beds of particles in Groups A and B at pressures of up to 25 bar showed that both bubbles and slugs broke up by fingers of particles falling in from the roof an effect that became more pronounced with increasing pressure.

5.4.2.2 Group B Materials

King and Harrison (1980) found bubble size to be independent of pressure up to 25 bar but Hoffmann and Yates (1986), also using X-rays, found mean bubble diameters to increase slightly up to 16 bar and to decrease thereafter up to 60 bar. This work also showed an increase in bubble coalescence as pressure was increased but that their stability was lower at higher pressures causing them to break up into ever smaller units; at the highest pressures studied bubbles were hard to identify at all the bed having taken on the appearance of an ill-defined foaming mass of fluidized material. These results were later confirmed in a study by Olowson and Almstedt (1990).

5.4.2.3 Group D Materials

These have been little studied relative to those in Groups A and B. King and Harrison (1980) studied spouted beds of 1.1 mm diameter glass spheres at pressures of up to 20 bar and found a marked decrease in minimum spouting velocity with increasing pressure and concluded that Group D materials should follow the same trends as those shown by Group B powders but at higher pressures.

5.4.3 *Jet Penetration*

When gas first enters a fluidized bed from an orifice in a supporting grid it does so either in the form of discrete bubbles or as a flame-like jet that decays into a stream of bubbles at some height above the grid. Whether jets or bubbles form was explored by Grace and Lim (1987) who, on the basis of much experimental evidence concluded that jets would form for values of the ratio:

$$\frac{d_{or}}{d_p} \leq 25.4 \tag{5.21}$$

where d_{or} and d_p are the diameters of the orifice and bed particles respectively. Under all other conditions bubbles rather than jets would form Hirsan et al. (1980) measured maximum jet penetration lengths, L_{max}, in beds of Group B materials up to pressures of 50 bar and found:

$$\frac{L_{max}}{d_p} = 26.6\left(\frac{\rho_f}{\rho_p}\right)^{0.67}\left(\frac{U_0^2}{gd_p}\right)^{0.34}\left(\frac{U}{U_{cf}}\right)^{-0.24} \tag{5.22}$$

where U_0 is the orifice gas velocity and U_{cf} is the superficial velocity necessary to fluidize the polydispersed powder. The correlation shows that jet penetration length increases with pressure but decreases as the velocity of the fluidizing gas increases. Other similar correlations have been obtained by Yang (1981) and Yates et al. (1986).

5.4.4 Entrainment and Elutriation

Entrainment occurs when gas bubbles burst at the bed surface and throw particles into the freeboard space. At low gas velocities these particles fall back to the bed surface and are retained but as fluidizing velocity increases more particles are transported to ever greater heights giving rise to a particle-density gradient above the surface. For sufficiently tall freeboards there will be a height at which the density gradient falls to zero and above this height the entrainment flux will be constant. This height is called the transport disengaging height or TDH. At sufficiently high gas velocities particles will be carried out of the bed completely or elutriated. Elutriation is considered to be a first-order process such that the rate of elutriation of particles within a size range d_{pi} is directly proportional to the mass fraction of that size range x_i in the bed. Thus:

$$-\frac{1}{A_t}\frac{\mathrm{d}}{\mathrm{d}t}(x_i M) = \kappa_i^* x_i \tag{5.23}$$

where A_t is the bed cross-sectional area, M is the mass of particles in the bed and κ_i^* is the elutriation rate constant with units kg/m^{2}s. From Eq. 5.23:

$$x_i = x_{i0}\exp\left(-\frac{\kappa_i^* A_t t}{M}\right) \tag{5.24}$$

where x_{i0} is the initial mass fraction of the particles at time zero. There are many empirical correlations for κ_i^* in terms of the physical properties of gas and particles (Kunii and Levenspiel 1991) from which it is clear that the particle terminal-fall velocity, u_t, is an important factor and that the rate coefficient increases as u_t decreases. Increasing pressure would thus be expected to increase the rate of elutriation, a result confirmed by Chan and Knowlton (1984) in a study of a bed of sand particles with a wide size distribution at pressures of up to 31 bar (Fig. 5.11)

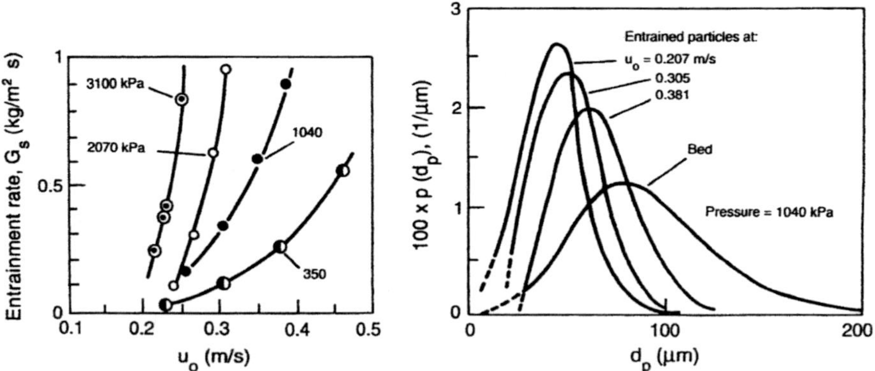

Fig. 5.11 Effect of pressure and fluidizing gas velocity on solids entrainment (Kunii and Levenspiel (1991) based on data of Chan and Knowlton (1984)

5.4.5 Heat Transfer

Fluidized-bed heat transfer is a major consideration in the design of reactors particularly those involving exothermic reactions. The field is very wide and has been reviewed in depth in a number of studies over the years a recent comprehensive example being that by Chen (2003) who pointed out that the mechanisms of transfer are significantly different for different fluidization regimes the two most important for industrial applications being bubbling beds and fast, circulating beds. Grace (1986) produced a "fluidization map" (Fig. 5.12) in which the various flow regimes are plotted as functions of a dimensionless particle diameter, d_p^*, and a dimensionless velocity, u^* where:

$$d_p^* = d_p \left[\frac{\rho_g (\rho_p - \rho_g) g}{\mu^2} \right]^{1/3} = Ar^{1/3} \tag{5.25}$$

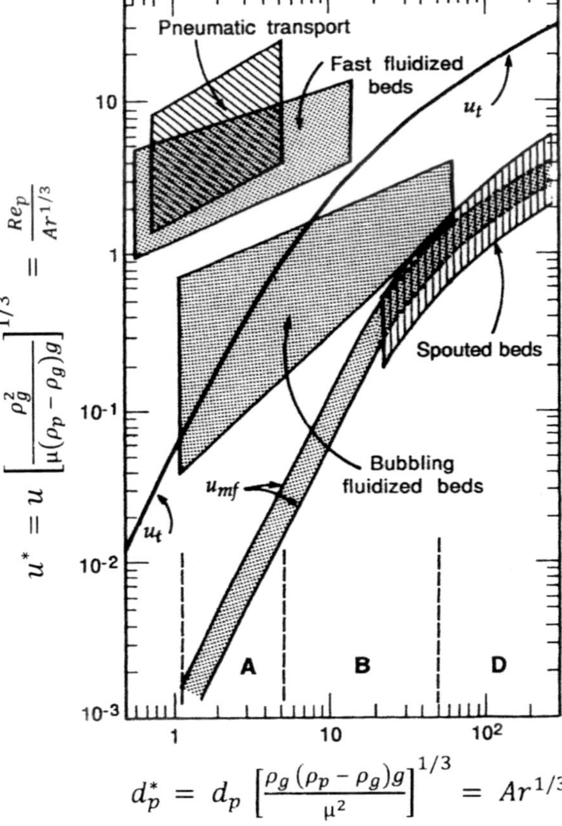

Fig. 5.12 Regions of operation of fluidized beds of industrial significance (After Grace 1986)

$$u^* = u\left[\frac{\rho_g^2}{\mu(\rho_p - \rho_g)g}\right]^{1/3} = \frac{Re_p}{Ar^{1/3}} \tag{5.26}$$

Incorporating the physical properties of particles and gas with values of fluidizing gas velocities thus enables the operating regime to be identified; most industrial reactors operate within the regions indicated on the map. Some of these applications involve operation at elevated pressures and it is important to appreciate how the transfer process changes from ambient as pressure is increased (Fig. 5.12).

5.4.5.1 Bubbling Beds

As a result of the high surface area available in the particulate phase heat transfer between fluidizing gas and bed particles is normally very efficient and will not be considered further here although it is treated in some detail in Chen (2003). It is generally accepted that the heat transfer coefficient between a bed and an immersed surface can be expressed as the sum of three components:

$$h = h_{pc} + h_{gc} + h_r \tag{5.27}$$

where h_{pc}, h_{gc} and h_r are the particle convective, gas convective and radiative transfer coefficients respectively. The gas convective term is of importance only for beds of large Group B and Group D materials while the radiative component is of significance only above about 600 °C so that for Group A and small Group B materials in the absence of radiation effects it is h_{pc} that dominates the heat transfer process. Several different approaches have been employed to estimate values of h_{pc}. Early work by Leva et al. (1949) on vertical surfaces proposed the following correlation for the heat-transfer Nusselt number in terms of k_g, the thermal conductivity of the fluidizing gas and Re_p the particle Reynolds number:

$$Nu_{pc} = \frac{h_{pc}d_p}{k_g} = 0.525(Re_p)^{0.75} \tag{5.28}$$

Similar correlations were proposed by Wender and Cooper (1958), Andeen and Glicksman (1976), Borodulya et al. (1991) and Molerus et al. (1995). An alternative approach, the so-called "packet theory" was originated by Mickley and Fairbanks (1955) and pictured bed-to-surface heat transfer as an unsteady-state process in which packets of emulsion-phase material carry heat to or from the surface, residing there for a short period of time before moving back into the bulk of the bed and being replaced by fresh material. The model gives a value for the instantaneous heat-transfer coefficient, h_i, as:

$$h_i = \left[\left(\frac{k_{mf}\,\rho_{mf}C_{mf}}{\pi\tau}\right)\right]^{1/2} \tag{5.29}$$

where k_{mf}, ρ_{mf} and C_{mf} are the thermal conductivity, density and heat capacity of the emulsion phase respectively and τ is the residence time of the packet at the wall. The effective conductivity of the emulsion phase may be given by:

$$k_{mf} = k_e^0 + 0.1\rho_g C_g d_p u_{mf} \tag{5.30}$$

k_e^0 being the conductivity of a fixed bed containing a stagnant gas. The main effect of increasing pressure will be to raise the gas density which will affect Re_p but have only a slight effect on h_{pc} through its influence on k_{mf}. For Group A and small Group B particles the suppression of bubbling caused by increasing pressure will increase the heat transfer by improving the quality of fluidization near the transfer surface as reported by Borodulya et al. (1982) who found an increase of 30 % in the maximum heat transfer coefficient for 0.126 mm sand particles between 6 and 81 bar. For larger particles the effect of pressure is to increase the gas-convective component via the increase in Re_{mf}. These trends have been confirmed by the work of Botterill and Desai (1972), Botterill and Denloye (1978), Staub and Canada (1987), Canada and McLaughlin (1978), Xavier et al. (1980) and Olsson and Almstedt (1995).

5.4.5.2 Circulating Beds

Owing to the danger of tube erosion caused by fast-moving solid particles heat exchange is normally carried out via cooling/heating tubes mounted in the walls of the vessel rather than by tubes immersed in the bed. The bed-to-wall heat exchange coefficient, h_w, is given by:

$$h_w = \frac{q}{a_w(T_b - T_w)} \tag{5.31}$$

where q is the rate of heat transfer, a_w is the area of exposed surface and T_b and T_w are the temperatures of bed and wall respectively. The many experimental measurements carried out to determine values of h_w have been reviewed by Chen (2003) who summarized the main observations to include:

- h_w is higher than that for gas convection at the same velocity but lower than that for bubbling beds
- h_w decreases with increasing particle size
- h_w increases with increasing solids mass flux
- h_w decreases with increasing height in the bed

There are many different correlations for h_w in the literature but no one is generally applicable and they will not be reviewed here—again the interested reader is referred to Chen (2003) for a comprehensive survey. One example will be given to illustrate the general approach. Werdermann and Werther (1993) proposed the following correlation for the particle-convective component of h_w for vertical surfaces in excess of 0.5 m in length in a CFB:

$$\frac{h_{pc}d_p}{k_g} = 7.46 \times 10^{-4}\left(\frac{D\rho_g U_g}{\mu_g}\right)^{0.757}\left(\frac{\rho_b}{\rho_p}\right)^{0.502}. \tag{5.32}$$

where D is the diameter of the column, U_g is the superficial gas velocity and ρ_b is the cross-sectional-average bed density (Grace and Bi 2003).

For circulating beds operated at high temperatures such as combustors the radiative component of h_w must be taken into account since these have been found to increase linearly with temperature to contribute over 35 % of the total at temperatures above 800°C (Ozkaynak et al. 1983). There are as yet few reports of heat-transfer measurements in circulating pressurized fluidized beds although their hydrodynamics have been studied in a number of cases. Thus Karri and Knowlton (1997) showed that solids hold-up decreased at pressures of 6.9 bar while Wirth and Gruber (1997) found solids to be more uniformly distributed over the full height of a CFB riser ay pressures of up to 50 bar (Grace and Bi 2003). Both effects would be expected to influence the heat-transfer performance of such units.

5.5 Conclusions

This chapter has demonstrated the important role that process conditions, namely temperature and pressure, play on the fluidization behavior of gas solid fluidized beds. Prediction of the fluidization behavior at process conditions is of major importance given that most of the industrial processes which use fluidized beds are operated at temperatures and pressures well above ambient. This chapter has explored the complexity of accounting for both hydrodynamic and interparticle effects with increasing temperature. It has also reviewed the effect of pressure on key design parameters such as entrainment, heat transfer and jet penetration. Although achieving a full understanding of the effect of process conditions on fluidization still remains a challenge, the theories and models presented in this chapter and developed over the last few decades have contributed to a key advancement in fluidized-bed design and operations.

References

Abrahamsen A, Geldart D (1980) Behaviour of gas-fluidized beds of fine powders part II. Voidage of the dense phase in bubbling beds. Powder Technol 26:47–55

Andeen BR, Glicksman LR (1976) Heat transfer to horizontal tube in shallow fluidized beds. Nat Heat Transfer Conf, St Louis MO, Paper 76-HT-67

Avidan AA, Yerushalmi J (1982) Bed expansion in high velocity fluidization. Powder Technol 32:223–232

Baerns M (1966) Effect of interparticle adhesive forces on fluidization of fine particles. Ind Eng Chem Fund 5(4):508–516

Boland D, Geldart D (1971) Electrostatic Charging in gas fluidized beds. Powder Technol 5:289–297

Borodulya VA, Epanov YG, Teplitskii YS (1982) Fluidized bed heat transfer. J Eng Phys 42:528

Borodulya VA, Teplitsky YS, Sorokin AP, Markevitch IL, Hassan AF, Yeromenko TP (1991) Heat transfer between a surface and a fluidized bed: considerations of temperature and pressure effects. Int J Heat Mass Transfer 34:47–53

Botterill JSM, Denloye AOO (1978) Ber-to-surface heat transfer in fluidized beds of large particles. Powder Tech 19:197–203

Botterill JSM, Desai M (1972) Limiting factors in gas-fluidized bed heat transfer. Powder Tech 6:231–238

Botterill JSM, Teoman Y, Yüregir KR (1982) The effect of operating temperature on the velocity of minimum fluidization, bed voidage and general behaviour. Powder Technol 31:101–110

Broadhurst TE, Becker HA (1975) Onset of fluidization and slugging in beds of uniform particles. AIChE J 21:238–247

Bruni G, Lettieri P, Newton D, Yates JG (2006) The influence of fines size distribution on the behaviour of gas fluidized beds at high temperature. Powder Technol 163:88–97

Canada GS, MacLaughlin MH (1978) Large-particle fluidization and heat transfer at high pressure. AIChE Symp Ser 74(176):27–37

Carman PC (1937) Fluid flow through granular beds. Trans Inst Chem Eng 15:150

Castellanos A (2005) The relationship between attractive interparticle forces and bulk behaviour in dry and uncharged fine powders. Adv Phys 54:263–376

Chan IH, Knowlton TM (1984) The effect of pressure on entrainment from bubbling gas-fluidized beds. In: Kunii D, Toei R (eds) Fluidization. Engineering Foundation, New York, pp 283–290

Chen JC (2003) Heat transfer. In: Yang W-C (ed) Handbook of fluidization and fluid-particle systems (Chapter 3). Marcel Dekker, New York

Chen ZD, Chen XP, Wu Y, Chen RC (2010) Study on minimum fluidization velocity at elevated temperature. Proc Chin Soc Electr Eng 30:21–25

Clift R, Grace JR, Weber ME (1974) Stability of bubbling fluidized beds. I E C Fundam 13:45–51

Compo P, Pfeffer R, Tardos GI (1987) Minimum sintering temperature and defluidization characteristics of fluidizable materials. Powder Technol 51:85–101

D'Amore M, Donsi' G, Massimilla L (1979) The influence of bed moisture on fluidization characteristics of fine powders. Powder Technol 23:253–259

Davidson JF, Harrison D (1963) Fluidized particles. Cambridge University Press, Cambridge

Doichev K, Akhmakov N (1979) Fluidisation of polydisperse systems. Chem Eng Sci 2–4

Donsì G, Massimilla L (1973). Bubble-free expansion of gas-fluidized beds of fine particles. AIChE J 19:1104–1110

Ennis BJ, Tardos G, Pfeffer R (1991) A microlevel-based characterisation of granulation phenomena. Powder Technol 65:257–272

Ergun S (1952) Fluid flow through packed columns. Chem Eng Prog 48(2):89–94

Fairbrother R (1999) A microscopic investigation of particle-particle interactions in the presence of liquid binders in relation to the mechanisms of "wet" agglomeration processes. PhD Dissertation, Department of Chemical Engineering, University College London

Formisani B, Girimonte R, Mancuso L (1998) Analysis of the fluidization process of particle beds at high temperature. Chem Eng Sci 53:951–961

Foscolo PU, Gibilaro LG (1984) A fully predictive criterion for the transition between particulate and aggregate fluidization. Chem Eng Sci 39(12):1667–1675

Foscolo PU, Gibilaro L (1987) Fluid dynamic stability of fluidised suspensions: the particle bed model. Chem Eng Sci 42(6):1489–1500

Geldart D, Wong ACY (1984) Fluidization of powders showing degrees of cohesiveness-I. Bed expansion. Chem Eng Sci 39(10):1481–1488

Geldart D, Wong ACY (1985) Fluidization of powders showing degrees of cohesiveness-II. Experiments on rates of de-aeration. Chem Eng Sci 40(4):653–661

Girimonte R, Formisani B (2009) The minimum bubbling velocity of fluidized beds operating at high temperature. Powder Technol 189:74–81

Girimonte R, Formisani B (2014) Effects of operating temperature on the bubble phase properties in fluidized beds of FCC particles. Powder Technol 262:14–21

Godard KMS, Richardson JF (1968) The behaviour of bubble-free fluidised beds. Inst Chem Eng Symp Ser 30:126–135

Goo JH, Seo MW, Kim SD, Song BH (2010) Effects of temperature and particle size on minimum fluidization and transport velocities in a dual fluidized bed. In: Proceedings of the 20th international conference on fluidized bed combustion. pp 305–310

Goroshko VD, Rozembaum RB, Toedes OH (1958) Approximate relationships for suspended beds and hindered fall. Izv Vuzov Neft Gaz 1:125

Grace JR (1986) Contacting modes and behaviour classification of gas-solid and other two-phase suspensions. Can J Chem Eng 64:353–363

Grace JR, Bi H (2003) Circulating fluidized beds. In: Yang W-C (ed) Handbook of fluidization and fluid-particle systems (Chapter 19). Marcel Dekker, New York

Grace JR, Lim CJ (1987) Permanent jet formation in beds of particulate solids. Can J Chem Eng 65(1):160–162

Grace JR, Sun G (1991) Influence of particle size distribution on the performance of fluidized bed reactors. Can J Chem Eng 69(5):1126–1134

Haider A, Levenspiel O (1989) Drag coefficient and terminal velocity of spherical and non-spherical particles. Powder Tech 58:63–70

Hamaker HC (1937) The London-Van der Waals attraction between spherical particles. Physica IV 10:1059–1068

Hartman M, Trnka O, Pohořelý M (2007) Minimum and terminal velocities in fluidization of particulate ceramsite at ambient and elevated temperature. Ind Eng Chem Res 46:7260–7266

Hirsan I, Shistla C, Knowlton TM (1980) The effect of bed and jet parameters on vertical jet penetration length in gas-fluidized beds. In: 73rd annual AIChE meeting, Chicago, Illinois

Hoffmann AC, Yates JG (1986) Experimental observations of fluidized beds at elevated pressures. Chem Eng Commun 41:133

Israelachvili J (1991) Intermolecular and surface forces. Academic Press, London

Jiliang M, Xiaoping C, Daoyin L (2013) Minimum fluidization velocity of particles with wide size distribution at high temperatures. Powder Technol 235:271–278

Karri SBR, Knowlton TM (1997) The effect of pressure on CFB riser hydrodynamics. In: Kwauk M, Li J (Eds) Circulating fluidized bed technology. Beijing Science Press, Beijing, pp 103–109

King DF, Harrison D (1980) The bubble phase in high-pressure fluidized beds. Grace JR, Matsen JM (eds) Fluidization. Plenum Press, New York, pp 101–107

King DH, Harrison D (1982) The dense phase of a fluidized bed at elevated pressures. Trans Inst Chem Eng 60:26–30

Knowlton TM (1992) Pressure and temperature effects in fluid-particle systems. Fluidization VII. Engineering Foundation, New York, pp 27–46

Kunii D, Levenspiel O (1991) Fluidization engineering. Butterworths, Boston

Landi G, Barletta D, Poletto M (2011) Modelling and experiments on the effect of air humidity on the flow properties of glass powders. Powder Technol 207:437–443

Landi G, Barletta D, Lettieri P, Poletto M (2012) Flow properties of moisturized powders in a Couette fluidized bed rheometer. Int J Chem Reactor Eng 10(A28):1–13

Lettieri P (1999) A study on the influence of temperature on the flow behaviour of solid materials in a gas fluidized bed. PhD thesis, University College, London

Lettieri P, Yates JG, Newton D (2000) The influence of interparticle forces on the fluidization behaviour of some industrial materials at high temperature. Powder Technol 110:117–127

Lettieri P, Newton D, Yates JG (2001a) High temperature effects on the dense phase properties of gas fluidized beds. Powder Technol 120:34–40

Lettieri P, Brandani S, Newton D, Yates JG (2001b) A generalization of the Foscolo and Gibilaro particle-bed model particle to predict the fluid stability of some fresh FCC catalysts at elevated temperatures. Chem Eng Sci 56(18):5401–5412

Leva M (1959) Fluidization. McGraw-Hill, New York

Leva M, Weintraub M, Grummer M (1949) Heat transmission through fluidized beds of fine particles. Chem Eng Prog 45:563–572

Leva M, Shirai T, Wen CY (1956) Prediction of onset of fluidization in beds of granular solids. Genie Chim 75:33

Lin C-L, Wey M-Y, You S-D (2002) The effect of particle size distribution on minimum fluidization velocity at high temperature. Powder Technol 126:297–301

Lucas A, Arnaldos J, Casal J, Puigjaner L (1986) High temperature incipient fluidization in mono and polydisperse systems. Chem Eng Commun 41:121–132

Massimilla L, Donsi' G (1976) Cohesive forces between particles of fluid-bed catalysts. Powder Technol 15(2):253–260

Massimilla L, Donsi' G, Zucchini C (1972) The structure of bubble-free gas fluidized beds of fine fluid cracking catalyst particles. Chem Eng Sci 27:2005–2015

Mickley HS, Fairbanks DF (1955) Mechanism of heat transfer to fluidized beds. AIChE J 1: 374–384

Miller C, Logwinuk A (1951) Fluidization studies of solid particles. Ind Eng Chem 43:1220–1226

Molerus O, Burschka A, Dietz S (1995) Particle migration at solid surfaces and heat transfer in bubbling fluidized beds II. Prediction of heat transfer in bubbling fluidized beds. Chem Eng Sci 50:879–885

Mutsers SMP, Rietema K (1977) The effect of interparticle forces on the expansion of a homogeneous gas-fluidized bed. Powder Technol 18:239–248

Newton D, Smith G, Hird N (1996) Assessment of FCC catalysts evaluation criteria. In: Presented at the fourth international conference on fluid particle interaction, Davos, Switzerland

Olowson PA, Almstedt AE (1990) Influence of pressure and fluidization velocity on the bubble behaviour and gas-flow distribution in a fluidized bed. Chem Eng Sci 45:1733–1741

Olowson PA, Almstedt AE (1991) Influence of pressure on the minimum fluidization velocity. Chem Eng Sci 46:637–640

Olsson SE, Almstedt AE (1995) Local instantaneous and time-averaged het transfer in a pressurized fluidized bed with horizontal tubes: influence of pressure, fluidization velocity and tube-bank geometry. Chem Eng Sci 50:3231–3245

Ozkaynak TF, Chen JC, Frankenfield TR (1983) An experimental investigation of radiant heat transfer in high-temperature fluidized bed. In: Fluidization. Engineering Foundation, New York, Vol V, pp 371–378

Raso G, D'Amore M, Formisani B, Lignola PG (1992) The influence of temperature on the properties of the particulate phase at incipient fluidization. Powder Technol 72:71–76

Riba JP, Routie R, Couderc JP (1978) Conditions minimales de miseenfluidisation par unliquide. Can J Chem Eng 56:26–30
Richardson J, Zaki WN (1954) Sedimentation and fluidisation: Part I. Trans Inst Chem Eng 32: 35–53
Rietema K, Piepers HW (1990) Effect of interparticle forces on the stability of gas-fluidized beds-I. Experimental evidence. Chem Eng Sci 45(6):1627–1639
Rietema K, Cottaar EJE, Piepers HW (1993) Effect of interparticle forces on the stability of gas-fluidized beds—II. Theoretical derivation of bed elasticity on the basis of van der Waals forces between powder particles. Chem Eng Sci 48(9):1687–1697
Rowe PN (1984) The effect of pressure on minimum fluidization velocity. Chem Eng Sci 39: 173–174
Rowe PN, Santoro L, Yates JG (1978) The division of gas between bubble and interstitial phases in fluidized beds of fine powders. Chem Eng Sci 33:133–140
Seville JPK, Clift R (1984) The effect of thin liquid layers on fluidization characteristics. Powder Technol 37:117–129
Seville JPK, Silomon-Pflug H, Knight PC (1998) Modelling of sintering in high temperature gas fluidization. Powder Technol 97(2):160–169
Siegell JH (1984) High-temperature defluidization. Powder Technol 38:13–22
Siegell JH (1989) Early studies of magnetized fluidized beds. Powder Technol 57:213–220
Simons SJR, Seville JPK, Adams MJ (1993) Mechanisms of agglomeration. In: Sixth international symposium on agglomeration, Nagoya, Japan
Staub FW, Canada GS (1987) Effect of tube bank and gas density on flow behaviour and heat transfer in a fluidized bed. In: Davidson JF, Kearns DL (eds) Fluidization. Cambridge University Press, Cambridge, pp 339–344
Subramani HJ, MothivelBalaiyya MB, Miranda LR (2007) Minimum fluidization velocity at elevated temperatures for Geldart's group-B powders. Exp Therm Fluid Sci 32:166–173
Tardos G, Mazzone D, Pfeffer R (1985) Destabilization of fluidized beds due to agglomeration, Part I: Theoretical model. Can J Chem Eng 63:377–383
Valverde JM, Castellanos A (2008) Bubbling suppression in fluidized beds of fine and ultrafine powders. Part Sci Technol 26:197–213
Valverde JM, Quintanilla M, Castellanos A, Mills P (2001) The settling of fine cohesive powders. Europhys Lett 54:329–334
Valverde JM, Castellanos A, Mills P, Quintanilla M (2003) Effect of particle size and interparticle force on the fluidization behavior of gas-fluidized beds. Phys Rev E Stat Nonlin Soft Matter Phys 67:051305
Wen CY, Yu YH (1966) Mechanics of fluidization. Chem Eng Progr Symp Ser 62:100–111
Wender L, Cooper GT (1958) Heat transfer between fluidized-solids bed and boundary surfaces—correlation of data. AIChE J 4:15–23
Werdermann CC, Werther J (1993) Solids flow pattern and heat transfer in an industrial-scale fluidized-bed heat exchanger. In: Proceedings of 12th international conference on fluid-bed combustion, vol 2, pp 985–990
Wirth KE, Gruber U (1997) Fluid mechanics of circulating fluidized beds with small-density ratio of solids to fluid. In: Kwauk M, Li J (eds) Circulating fluidized-bed technology. Beijing Science Press, Beijing, pp 78–83
Wu S, Baeyens J (1991) Effect of operating temperature on minimum fluidization velocity. Powder Technol 67:217–220
Xavier AM, King DF, Davidson JF, Harrison D (1980) Surface-bed heat transfer in a fluidized bed at high pressure. In: Grace JR, Matsen JM (eds) Fluidization. Plenum, New York, pp 209–216
Xie HY, Geldart D (1995) Fluidization of FCC powders in the bubble-free regime: effect of types of gases and temperature. Powder Technol 82:269–277

Yamazaki R, Han NS, Sun ZF, Jimbo G (1995) Effect of chemisorbed water on bed voidage of high temperature fluidized bed. Powder Technol 84:15–22

Yang W-C (1981) Jet penetration in a pressurized fluidized bed. I E C Fundam 20:297–300

Yates JG (1996) Effects of temperature and pressure on gas fluidization. Chem Eng Sci 51: 167–205

Yates JG (2003) Effect of temperature and pressure. In: Yang W-C (ed) Handbook of fluidization and fluid-particle systems (Chapter 5). Marcel Dekker, New York

Yates JG, Bejcek V, Cheesman DJ (1986) Jet penetration into fluidized beds at elevated pressures. In: Ostergaard K, Sorensen S (eds) Fluidization V. Engineering Foundation, New York, pp 79–86

Chapter 6
Fluidized-Bed Scaling

Abstract The principles on which the performance of a full-scale fluidized-bed reactor may be inferred from that of a cold, scaled-down model are outlined and lead to a review of the scaling rules developed in the recent past. Dimensional analysis based on the Buckingham π-theorem is described as well as the alternative approach based on the governing equations of conservation of mass and momentum of fluidized particles. Examples are given of both rigorous and simplified sets of dimensionless groups appropriate to the scaling process and a description is given of the way they are applied to bubbling beds. This is followed by a consideration of the scaling relationships relevant to circulating fluidized-bed combustors where additional groups such as the Damköhler numbers can be applied. Work on the validation of the scaling rules is then described and leads to a section in which scaling is analysed in terms of the non-linear chaotic behaviour of fluidized beds. The chapter ends with a description of the application of the scaling rules to a scaled-down model of a thermal denitration reactor and its internal structure as revealed by X-ray analysis.

6.1 Introduction

The development of a new process centred on a fluidized-bed reactor proceeds through a number of stages. Experiments on a laboratory bench-scale unit provide basic information concerning reaction kinetics, catalyst activity and deactivation, particle attrition and agglomeration etc. This stage would typically be followed by work on larger scale pilot and demonstration units proceeding eventually to the full-scale plant. In the early stages of development a decision would be made regarding the type of fluidized bed to be used: bubbling, turbulent, circulating, entrained flow etc. and information would be needed on the expected hydrodynamic behaviour of the chosen system. It would be desirable to be able to infer the behaviour of the full-scale unit from that of, say, the laboratory unit or the pilot plant i.e. to be able to scale up the hydrodynamics from the smaller to the bigger bed. In practice this is far from straightforward since large beds have different solids

J.G. Yates and P. Lettieri, *Fluidized-Bed Reactors: Processes and Operating Conditions*, Particle Technology Series 26,
DOI 10.1007/978-3-319-39593-7_6

circulation and gas-solid contacting patterns from smaller beds and a unit that performs well at pilot scale often falls short of expectation in the full-scale plant (Fitzgerald et al. 1984). To address this issue much work has been reported on efforts to develop criteria for hydrodynamic similarity between fluidized beds of different scales, temperatures and pressures and to identify the relevant parameters and variables necessary to achieve dynamic similarity. To this end it is necessary to match certain dimensionless groups that must be kept equal at all scales, a procedure traditionally used in other areas of engineering such as aircraft and ship design where wind tunnels and flow tanks are used to explore fluid flows, drag forces, pressure profiles etc. around small-scale models. To identify the relevant scaling parameters for fluidized beds dimensional analysis may be applied via the Buckingham π-theorem or by analysis of the governing differential equations and boundary conditions that completely define the system under consideration. A prerequisite of any such analysis is that the units to be matched by hydrodynamics must be geometrically similar i.e. they must be the same shape and all their linear dimensions must be related by a constant scale factor.

6.2 Dimensional Analysis

An early method is based on the Buckingham π-theorem. This states that the n independent parameters defining any physical system may be reduced to (n-k) dimensionless groups where k is the number of dimensionally independent parameters whose value is less than or equal to the number of dimensions (mass (M), length (L) and time (T)) in the original defining n parameters. Glicksman et al. (1994) demonstrated the use of the theorem to determine the dimensionless groups that govern the hydrodynamic behaviour of fluidized beds. Expressing the pressure drop through a bed, ΔP, as the dependent parameter in terms of the main independent parameters:

$$\Delta P = f(u_0, g, D, L, d_p, \rho_s, \rho_f, \mu, \phi) \tag{6.1}$$

where:

Symbol	Definition	Dimensions
u_0	superficial velocity	L/T
g	acceleration due to gravity	L/T^2
D	bed diameter	L
L	bed height	L
d_p	particle diameter	L
ρ_s	solids density	M/L^3
ρ_f	fluid density	M/L^3

(continued)

(continued)

Symbol	Definition	Dimensions
μ	fluid viscosity	M/(LT)
ϕ	particle sphericity	–

Choosing u_0, D and ρ_f as the dimensionally independent parameters and using these to non-dimensionalize the remainder leads to the following set of parameters that define the system:

$$\frac{\Delta P}{\rho_f u_0^2} = f\left(\frac{gD}{u_0^2}, \frac{L}{D}, \frac{d_p}{D}, \frac{\rho_s}{\rho_f}, \frac{\mu}{\rho_f u_0 D}, \phi\right) \tag{6.2}$$

As noted by Rüdisüli et al. (2012) however the π-theorem does not indicate whether the chosen list of independent parameters is complete, a problem not found with the alternative approach based on the governing equations of conservation of mass and momentum of fluidized particles. Several groups have explored this area (Horio et al. 1986; Zhang and Yang 1987; Foscolo et al. 1990; Chan and Louge 1992) but the most comprehensive investigations are those of Glicksman's group at MIT summarised by Glicksman et al. (1994) and Glicksman (2003). On the basis of the conservation equations of Anderson and Jackson (1967) they derived a so-called "full" set of dimensionless parameters as follows:

$$\frac{\rho_f \rho_s d_p^3 g}{\mu^2}, \frac{\rho_s}{\rho_f}, \frac{u_0^2}{gD}, \frac{\rho_{f u_0 D}}{\mu}, \frac{G_s}{\rho_s u_0}, \phi \tag{6.3}$$

For similarity it is also necessary to match the particle size distribution of the fluidized materials in both systems. In (6.3) the first term is the Archimedes number Ar (the ratio of gravitational to viscous forces), the third term is the Froude number Fr (the ratio of inertial to gravitational forces), the fourth is the Reynolds number Re_p(the ratio of inertial to viscous forces) and the fifth is the dimensionless solids circulation flux where G_s is the solids mass flux (kg/m^{2}s); this latter term is only of relevance for circulating beds.

Calculation of the operating conditions and parameter values for a large-scale bubbling-bed combustor and a small-scale cold model using air at standard conditions and based on (6.3) have been set out by Fitzgerald et al. (1984) and by Glicksman et al. (1994) as follows.

The gas/solid density ratios for the model (subscript m) and the combustor (subscript c) are matched as:

$$\left(\frac{\rho_f}{\rho_s}\right)_m = \left(\frac{\rho_f}{\rho_s}\right)_c \tag{6.4}$$

The Reynolds number and Froude number may be combined to give:

$$\frac{\rho_f u_0 D}{\mu}\frac{(gD)^{\frac{1}{2}}}{u_0} = \left(\frac{D^{\frac{3}{2}}}{\nu}g^{\frac{1}{2}}\right)_m = \left(\frac{D^{\frac{3}{2}}}{\nu}g^{\frac{1}{2}}\right)_c \tag{6.5}$$

where ν is the kinematic viscosity of the fluidizing gas. From (6.5):

$$\left(\frac{D_m}{D_c}\right) = \left(\frac{\nu_m}{\nu_c}\right)^{\frac{2}{3}} \tag{6.6}$$

It is further shown that:

$$\left(\frac{D}{d_p}\right)_m = \left(\frac{D}{d_p}\right)_c \tag{6.7}$$

and:

$$\frac{u_{0m}}{u_{0c}} = \left(\frac{\nu_m}{\nu_c}\right)^{\frac{1}{3}} = \left(\frac{D_m}{D_c}\right)^{\frac{1}{2}} \tag{6.8}$$

Satisfying (6.6) and (6.8) the Reynolds and Froude numbers are kept identical.

Based on these relationships a comparison between a hot combustor and a cold model is shown in Table 6.1 while the relevant values for a pressurised combustor are shown in Table 6.2. In the latter case the two units are comparable in size but a reduction in model dimensions could be achieved by use of a gas of higher density than air such as a Freon.

As this shows in practice it is sometimes difficult to match all the groups between a large hot reactor such as a combustor and a cold model of

Table 6.1 Atmospheric combustor modelled by a bed fluidized with air at ambient conditions (Glicksman et al. 1994)

Given	Commercial bed	Scale model
Temperature (°C)	850	25
Gas viscosity (10^{-5} kg/ms)	4.45	1.81
Density (kg/m^3)	0.314	1.20
Derived from scaling laws		
Solid density	ρ_{sc}	$3.82\rho_{sc}$
Bed diameter, length etc.	D_c	$0.225D_c$
Particle diameter	d_{pc}	$0.225d_{pc}$
Superficial velocity	u_{0c}	$0.47u_{0c}$
Volumetric solid flux	$(G_s/\rho_s)_c$	$0.47(G_s/\rho_s)_c$
Time	t_c	$0.47t_c$
Frequency	f_c	$2.13f_c$

Table 6.2 Pressurized combustor modelled by a bed fluidized with air at ambient conditions (Glicksman et al. 1994)

Given	Commercial bed	Scale model
Temperature (°C)	850	25
Gas viscosity (10^{-5} kg/ms	4.45	1.81
Density (kg/m^3)	3.14	1.20
Pressure (bar)	10	1
Derived from scaling laws		
Solid density	ρ_{sc}	$0.382\rho_{sc}$
Bed diameter, length etc	D_c	$1.05D_c$
Particle diameter	d_{pc}	$1.05d_{pc}$
Superficial velocity	u_{0c}	$1.01u_{0c}$
Volumetric solid flux	$(G_s/\rho_s)_c$	$1.01(G_s/\rho_s)_c$
Time	t_c	$1.01t_c$
Frequency	f_c	$0.98f_c$

laboratory-scale dimensions. Rüdisüli et al. (2012) cite the example of a hot reactor 1.60 m in diameter operated at 320 °C and 2.5 bar pressure. To scale this unit with a cold model fluidized by air at ambient conditions requires a bed of 1.48 m diameter. Scaling down to 0.2 m diameter would require the use of particles of density 23,000 kg/m^3 operated at a pressure of 20 bar. To overcome this problem Glicksman et al. (1993) sought to relax some of the criteria on which the full set of scaling groups were based and so to reduce the number required for similarity. This was achieved by modifying the form of the fluid- and particle-phase stress tensors in the basic equations of motion at the viscous and inertial limits represented by the Ergun equation (6.9). This expresses the pressure drop, ΔP, through a bed of particles with a voidage ε as:

$$\frac{\Delta P}{L} = \frac{150(1-\varepsilon)^2}{\varepsilon^3}\frac{\mu u_0}{(\emptyset d_p)^2} + \frac{1.75(1-\varepsilon)}{\varepsilon^3}\frac{\rho_f u_0^2}{\emptyset d_p} \tag{6.9}$$

The first term on the right-hand side represents the pressure loss due to viscous effects while the second term accounts for the effects of inertia. For flow through fine particles at low Reynolds numbers ($\mathrm{Re_p} < 4$) the viscous term dominates while for large particles and high Reynolds numbers ($\mathrm{Re_p} > 1000$) the inertial term is dominant. For the viscous limit the governing parameters were shown to be:

$$\frac{\rho_s u_0 d_p^2}{\mu D}, \frac{gD}{u_0^2}, \frac{D}{L}, \phi \tag{6.10}$$

The product of the first and second terms was shown (Glicksman 1988) to be equivalent to the ratio of the superficial velocity, u_0, and the minimum fluidization velocity, u_{mf}, thereby removing dependence on the Archimedes number and making the governing list:

$$\frac{u_{mf}}{u_0}, \frac{gD}{u_0^2}, \frac{D}{L}, \phi \tag{6.11}$$

In the inertial limit the governing list is:

$$\frac{gD}{u_0^2}, \frac{\rho_f}{\rho_s}, \frac{d_p}{D}, \frac{L}{D}, \phi \tag{6.12}$$

Glicksman et al. (1994) then combined (6.11) and (6.12) to give a set of parameters approximately valid for the intermediate region:

$$\frac{u_{mf}}{u_0}, \frac{gD}{u_0^2}, \frac{\rho_f}{\rho_s}, \frac{L}{D}, \phi \tag{6.13}$$

The advantage of the simplified set of scaling parameters is that they allow greater flexibility in the choice of dimensions of the small-scale unit removing the need for "exotic" particles and pressures (Rüdisüli et al. (2012).

Nicastro and Glicksman (1984) tested experimentally the full set of scaling parameters (6.3) by comparing the performance of a 0.61 m^2 and 4.4 m tall fluidized-bed combustor operated at 780 °C with a quarter-scale cold model. The operating conditions of the two beds are shown in Table 6.3 from which it may be seen that there is good agreement for all parameters except the density ratio as a result of the density of the iron powder being somewhat too low.

The similarity between the beds was tested by comparing pressure signals measured at different locations in the two beds. Figure 6.1 shows the power spectral

Table 6.3 Operating conditions in a coal-burning combustor and cold scale models fluidized with air (Nicastro and Glicksman 1984)

Bed material	Hot bed	Cold bed	Cold bed
	Sand and coal	Iron grit	Sand and coal
T_b (K)	1098	300	299
ρ_s (kg/m^3)	2630	7380	2630
d_p (μm)	677	170	677
u_0 (m/s)	0.93	0.47	0.94
L_f (m)	0.92	0.24	0.23
ε_{mf}	0.49	0.57	0.49
ε_f	0.60	0.64	0.62
u_{mf}	0.16	0.10	0.18
Re_p	5.17	5.33	41.8
Fr	129	130	132
ρ_s/ρ_f	7280	5920	2170
L_f/d_p	1360	1410	330
L/L_f	0.66	0.64	0.67

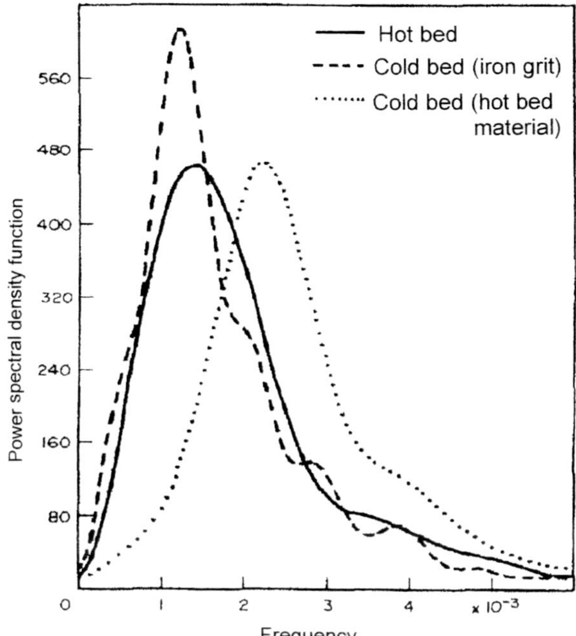

Fig. 6.1 Comparison of dimensionless power spectra of differential pressure fluctuations (Nicastro and Glicksman 1984)

density function and frequency measured by a probe situated at the wall. The plot shows good agreement between the hot bed and the cold bed containing iron powder but little agreement between the hot bed and the cold sand-containing bed indicating that modelling with identical bed material in a geometrically similar cold unit does not give dynamic similarity.

Roy and Davidson (1988) also used pressure measurements to compare bubbling beds at different temperatures and pressures—identities of dimensionless frequency and amplitude of pressure fluctuations indicating similarity. Their results showed that in the viscous limit at $Re_p < 30$ the reduced set of parameters given in 6.5/6.6 is sufficient to ensure dynamic similarity but that the full set (6.3) is necessary at $Re_p > 30$.

A number of other authors have developed scaling parameters for bubbling beds based on principles similar to or different from the above. Fitzgerald et al. (1984) also used the analysis of Anderson and Jackson (1967) to derive four dimensionless groups for similarity: the Froude number, the particle Reynolds number, the gas-to-solid density ratio and the ratio of a characteristic bed dimension to the average particle size. They used pressure fluctuation measurements to compare four different beds one of which was a 1.83 m^2 atmospheric combustor and another 0.46 m^2 cold bed of copper particles fluidized with air; autocorrelation plots of the pressure fluctuations for the two were found to be of the correct scaled frequency. Based on phenomenological models of bubble splitting and coalescence Horio et al. (1986) derived two scaling parameters:

$$\frac{u_0 - u_{mf}}{(gD)^{1/2}}, \frac{u_{mf}}{(gD)^{1/2}} \tag{6.14}$$

which were shown by Glicksman (1988) to be equivalent to those in (6.11) and thus to be only valid at the viscous limit. In a subsequent development by Horio et al. (1989) scaling relationships for circulating fluidized beds were obtained based on a model that assumes the riser of a CFB to have a core/annular structure with clusters of particles moving upwards in the core region and downwards in the annular region at the walls. Scaling was based on the equality of voidage distribution, dimensionless core radius, gas and solids splitting between core and annulus and cluster voidage. The similarity rules were tested experimentally using two geometrically similar scaled models (1/25 and 1/100) of a 175 MW CFB combustor. Axial voidage distribution, its transition and the radial distribution of cluster velocity in the scaled units were found to be in good agreement with those in the full-scale unit showing the validity of the proposed scaling law. Chang and Louge (1992) also considered scaling relationships for circulating fluidized beds on the basis of the continuum equations referred to above, deriving five dimensionless groups for flow of spherical particles in risers similar to those given in (6.3). For non-spherical materials however they followed the Ergun equation in combining the sphericity, ϕ, with the particle diameter, d_p, to produce the following:

$$\frac{\rho_f \rho_s (d_p \phi^\alpha g)^3}{\mu^2}, \frac{\rho_s}{\rho_f}, \frac{u_0}{(g d_p \phi^\alpha)^{\frac{1}{2}}}, \frac{D}{d_p \phi^\alpha}, \frac{G_s}{\rho_s u_0} \tag{6.15}$$

Here d_p is the diameter of a sphere having the same volume as the non-spherical particle, α is an empirical constant and D is the diameter of the riser. Experimental tests were carried out with three risers 0.32, 0.46 and 1 m in diameter with plastic, glass and steel powders, static pressures and pressure fluctuations being used to test for dynamic similarity. The results showed that in risers of moderate diameter vertical pressure profiles scale with riser diameter and particle density whereas pressure fluctuations scale with the product of particle diameter, density and sphericity.

On the basis of the one-dimensional "particle-bed model" Foscolo et al. (1990) derived a further set of scaling parameters:

$$\frac{g d_p^3 \rho_f^2}{\mu^2}, \frac{\rho_f}{\rho_s}, \frac{u_0}{u_t} \tag{6.16}$$

where u_t is the terminal-fall velocity of a single particle. They used these to compare the observed behaviour of a number of fluidized systems comprising different solids and fluids e.g. alumina—high pressure CF_4 gas (Crowther and Whitehead 1978), copper-water (Gibilaro et al. 1986), carbon-synthesis gas (Jacob and Weimer 1987). The object was to compare the bed voidage, ε_{mb}, at the point of transition from homogeneous to bubbling fluidization. Thus a copper-water system was matched to a carbon-high pressure gas (124 bar) system when both showed closely similar ε_{mb} values of 0.66 and 0.68 respectively. A system of soda-glass particles fluidized with water was matched with one of alumina particles fluidized by high-pressure gas: both, as predicted by the model, showing stable homogeneous behaviour throughout the range of velocities investigated (Gibilaro 2001).

6.3 Combustion Scaling

Leckner et al. (2011) reviewed work on the scale-up of circulating fluidized-bed combustors and described research carried out at Chalmers University with a 12 MW boiler and a 1/9th scale plastic model operated at ambient conditions. The boiler was operated with sand and low-ash wood chips while the scaled unit used iron and steel particles whose densities and shapes deviated somewhat from the required values indicated by the scaling set (6.3). The authors emphasised the fundamental difficulty in dynamic scaling of finding particles of the correct size, shape and density for the cold model. Experimental results in the case of the iron particles showed good correspondence between the solids volume fraction along the height of the riser in both model and boiler; correspondence in the case of the steel particles was not so good and was unexplained. Good agreement between gas-dispersion measurements was found for both materials as shown in Fig. 6.2.

The authors then discussed the question of combustion scaling as opposed to hydrodynamic scaling. In the former case combustion usually takes place in both the small and large plants and for scaling purposes a number of parameters may be

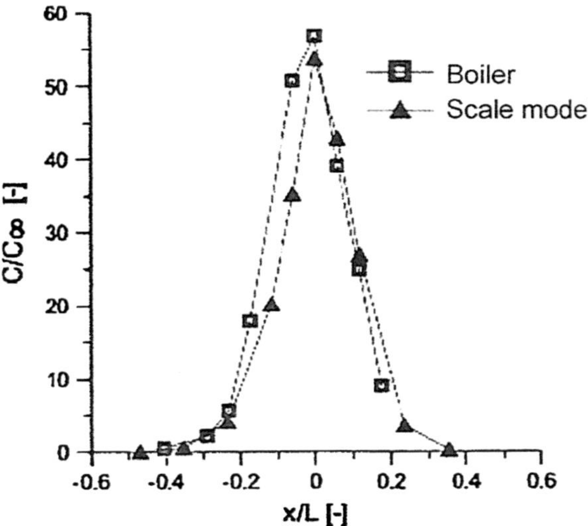

Fig. 6.2 Gas-concentration profiles from tracer-gas injection in the boiler and scale model (Leckner et al. 2011)

maintained identical in a test plant and a full-scale boiler. Such parameters are bed temperature, total excess-air ratio, primary-air stoichiometry, fuel and bed material. The fluidization-gas velocity should also be kept to the same order of magnitude in the two plants. Scaling then requires determination of the linear dimensions (height, L, and diameter, D) of the test riser as well as that of the solids mass flux. In general it is only feasible to apply scaling criteria to the riser of a CFB boiler since in a small-scale unit, although the riser height may be of comparable size, the diameter must be kept small; normally L/D > 30 in the model and L/D < 10 in the boiler. Leckner et al. (2011) firstly considered the horizontal-scaling problem related to the transport of fuel particles from distribution points at or near the wall of a riser. They cited the earlier work of Leckner and Werther (2000) who proposed the Damköhler number *Da* (ratio of transport time to reaction time) as a criterion for combustion scaling. Values of *Da* were determined on the basis of the two processes occurring when fuel is introduced into a combustor namely devolatilization and char combustion. The devolatilization/reaction time, t_v, for a particle of diameter d_p was taken as:

$$t_v = ad_p^2 \tag{6.17}$$

where a, an empirical constant, was given the value 10^6 s/m^2. Char combustion was assumed to be diffusion controlled giving the burn-out time, t_c, as:

$$t_c = \frac{\rho_c d_p^2}{48ShD_g c_0} \tag{6.18}$$

The average dispersion distance, x, was determined from an expression derived by Einstein (Gardiner 1997) for the dispersion time, t_d, in Brownian motion:

$$t_d = x^2/2D_h \tag{6.19}$$

where the horizontal dispersion coefficient, D_h, had a value 0.01 m^2/s. For high-volatile fuels reaction time was equated to devolatilization time and the horizontal Damköhler number, Da_h, results from combining (6.17) and (6.19) to give:

$$Da_h = \left(\frac{x}{d_p}\right)^2 \left(\frac{1}{2aD_h}\right) \tag{6.20}$$

$$= 50\left(\frac{x}{1000d_p}\right)^2 \tag{6.21}$$

and hence $Da_h \leq 1$ for $(x/1000d_\mathrm{p}) \leq 0.14$. This result shows that for a 1 mm diameter high-volatile fuel reaction in the horizontal direction will be completed in risers of diameter 0.14 m or less and for a 10 mm diameter material 1.4 m or less. The consequence is that scale-up from a small to a larger unit will be unreliable

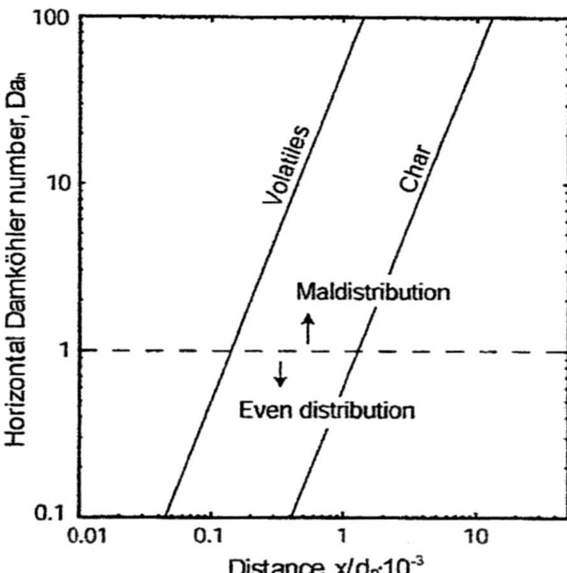

Fig. 6.3 Horizontal Damköhler number versus dimensionless dispersion distance (Leckner et al. 2011)

since maldistribution of fuel and air will be prevalent in a large plant but not in a small-scale unit where mixing will be more efficient. For char combustion Da_h is formed from (6.17) and (6.18) leading to $Da_h \leq 1$ for (x/1000d_p ≤ 1.4 and so fuel dispersion should be at less than 1.4 m for a 1 mm char particle and scale-up should be reliable. These results are shown diagrammatically in Fig. 6.3 which indicates that devolatilization is much faster than char combustion and that Da_h depends on the size of the fuel particle: larger particles are transported further before undergoing reaction.

These conclusions were borne out by the report by Alliston and Wu (1996) on work with a small-scale combustor burning bituminous coal in a bed of limestone and a 5 m diameter combustor where the pilot plant always performed better in terms of sulphur capture than the larger bed. Mixing was less critical in the smaller bed where the fuel-air mixture was more homogeneous in the vicinity of the fuel injection point.

Scaling in the vertical direction of a riser was again a function of fuel-particle size and composition. If fuel-air mixing at the entry point is efficient volatiles combustion will be complete before particles exit to the cyclone. Char combustion however is slower and a fraction of the material $(1 - \eta)$ will leave the cyclone and need to be recirculated; here $(\eta < 1)$ is the cyclone collection efficiency. The time spent by char particles in the reactive environment of the riser of length L is then:

$$t_t = \frac{L/u_p}{(\eta - 1)} \tag{6.22}$$

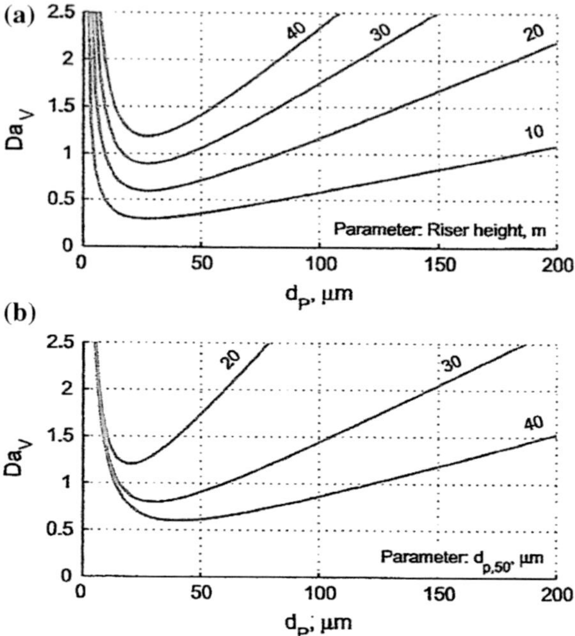

Fig. 6.4 Vertical Damköhler number versus char particle size (Leckner et al. 2011)

where $u_p = (u_0 - u_t)$, u_t being the terminal fall velocity of a single particle. The vertical Damköhler number is then:

$$Da_v = t_t / t_{char} \tag{6.23}$$

For vertical scaling Da_v should be greater than unity and as shown in Fig. 6.4a small particles will have sufficient time to react in their passage through tall risers; the influence of the cyclone efficiency on char burn-out is shown in Fig. 6.4b.

6.4 Validation of the Scaling Laws

Differential pressure fluctuations as measured by pressure probes immersed in the bed have frequently been used to test the validity of the scaling relationships. The fluctuations may be analysed statistically in terms of their spectral power density, as was demonstrated by Fitzgerald et al. (1984) and Nicastro and Glicksman (1984), or their probability density function as used by Sanderson and Rhodes (2005). In this latter study the simplified laws of Glicksman and Horio were tested with a set of four cylindrical cold model beds ranging in diameter from 146 to 1560 mm i.e. a ten-fold difference; the beds were operated at ambient temperature with spherical glass particles of various sizes fluidized with air. Pressure fluctuation measurements were made with probes situated at a number of locations distributed axially and radially within the beds. Based on the statistics derived from these measurements the authors generated an "agreement map" showing the extent of agreement with the scaling parameters in various regions of the beds. Good agreement was found generally with the small-scale beds but with the largest bed at gas velocities up to 3.5 u_{mf} poor agreement was found at the walls and towards the bed surface.

Di Felice et al. (1992) studied the scale-up rules established earlier (6.16) using five different gas-solid systems fluidized with air at ambient temperature and at different pressures in the bubbling and slugging regimes. Three of the systems examined used spherical particles and were dynamically similar, one used non-spherical particles and one was deliberately mismatched; again pressure fluctuations were applied to test similarities. Good agreement was found for the dynamically similar systems in the bubbling regime but not when the beds were in operated in slugging mode. Poor agreement was found for the other two systems.

Chaos analysis

A fluidized bed can be considered to be a non-linear chaotic system in which the governing variables may be projected into a multi-dimensional state space represented by a so-called "attractor" which gives a characteristic fingerprint of the system (van Ommen et al. 1999). Beds showing similar attractors may be considered to have similar hydrodynamic properties whereas variations in one bed's attractor over time indicate changes in its hydrodynamics. Such changes with time have been used to detect particle agglomeration (van Ommen et al. 2000). It was shown by Takens (1981) that the attractor may be reconstructed from the time series of one characteristic variable such as the local pressure variation (Fig. 6.5) and to compute the attractor of a fluidized bed B_1 a series of instantaneous pressure measurements ($p_1, p_2 \ldots p_N$) is made.

The values are then normalised by subtracting their average value from each reading and dividing by the standard deviation to give a time series x_k with N values a mean of zero and a standard deviation of unity. A similar procedure is followed for a second bed B_2 to give a time series y_k. The pressure-time series x_k is then converted to a set of $(N - m + 1)$ delay vectors X_k with m elements which can be considered as points on an m-dimensional state space leading to the reconstructed attractor for B_1 denoted as $\rho_X(X_i)$. The attractor for B_2, ($\rho_Y(Y_i)$), is constructed in a

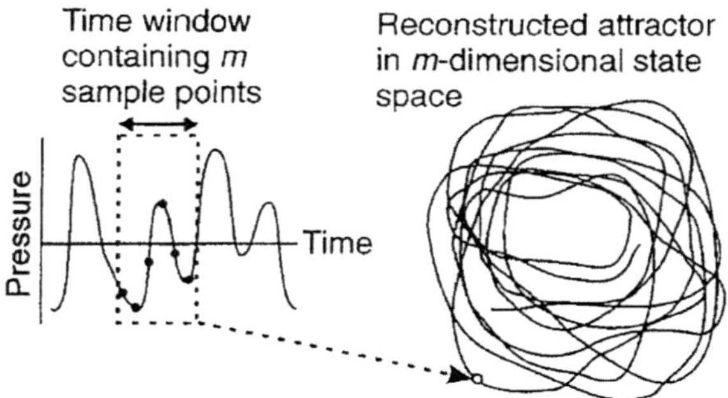

Fig. 6.5 Reconstruction of an attractor in the m-dimensional state space from a pressure-time series (van Ommen et al. 2000)

similar way and the extent to which the two attractors differ is given by the squared distance Q between them. An unbiased estimator Q' of this difference was calculated by Diks et al. (1996) and this along with the variance of the estimator V leads to a defining statistic S such that:

$$S = \frac{Q'}{\sqrt{V(Q')}} \tag{6.24}$$

If S is close to zero the two attractors and hence the two hydrodynamics are similar and the beds are correctly scaled; if $S > 3$ the two are different.

The attractor-comparison method was used by van Ommen et al. (2004) in an attempt to validate the scaling rules proposed by Horio et al. (1986). They found that while the original tools usually indicated similarity the statistical method showed disagreement. A similar conclusion was reached in a separate study by van Ommen et al. (2006). Pending further investigations the question remains open.

6.5 Application of the Scaling Laws to the Thermal Denitration Reactor at Sellafield, UK

In this section we present an example of the application of the scaling rules developed by Glicksman et al. (1994) applied to the design of a 4/10 scaled down model of the Magnox Reprocessing Thermal Denitration (TDN) fluidized-bed reactor operated at the nuclear fuel reprocessing site at Sellafield. This application is an example where employing in a laboratory scale model the fluids and the real solids as in the commercial process (and the reaction temperature), is impossible. Hence, in this case the scaling rules for fluidization are a fundamental tool to guide the design of a scale down system in which the fluid-dynamics of the real scale reactor may be replicated.

The Magnox reprocessing and uranium finishing plants have been at the heart of the UK's nuclear fuel reprocessing programme for over 50 years, reprocessing over 50,000 tonnes of irradiated uranium fuel from the UK's fleet of Magnox nuclear power stations. The Uranium Finishing Line, principally the Thermal Denitration Reactors (TDN's) convert uranyl nitrate (liquid) into uranium trioxide (a solid powder product) that can be manufactured into fuel and re-introduced into the nuclear fuel cycle. The TDN reactor is a fluidized-bed reactor, in which heated fluidizing air is introduced through nozzles at the base of the reactor to thermally de-nitrate the uranyl nitrate forming uranium trioxide. The aging MagnoxTDN reactors were becoming increasingly unreliable and restricted throughput of the reprocessing plant on many occasions. A major project was therefore initiated at UCL (Lettieri et al. 2014; Materazzi and Lettieri 2016) in collaboration with the National Nuclear Laboratory (NNL) and Sellafield Ltd. to investigate and resolve a number of operational problems occurring in the full scale TDN reactor at

Sellafield. The dimensionless parameters proposed by Glicksman et al. (1994) proved to be a reliable guide for the design of a 4/10 scale lab model operated under ambient conditions at UCL.

In this project, the unique X-ray imaging facility available at UCL (Lettieri and Yates 2013) was used to reveal for the first time the flow patterns inside such reactors and their fluidization performance (Holmes et al. 2015). X-ray studies of full scale sections of commercial units have been used to assess proposed process improvements, particularly in cases where there are significant internal hardware components such as cooling/heating coils, liquid spray nozzles, and feed gas spargers (Newton 2004). The advantage of X-raying a full scale section is that any uncertainties about the experimental conclusions are minimised even in cases where no reactions are taking place in the model reactor.

The process design followed to realize the 4/10 scaled-down TDN reactor is described: prior to the application of the scaling rules for fluidization, the geometry of the commercial reactor had to be scaled down. In this case a 4/10th scale was identified with NNL as being sufficiently large to avoid undesired interference from the reactor walls, whilst also allowing X-ray examination of the vessel. The complex reactor geometrical configuration of the real scale reactor and the hydrodynamic parameters were maintained in the scale down model to reproduce the fluidisation behaviour under ambient conditions. Hence, the geometric configuration for the bed, i.e. height-to-width ratio, internals (heating tubes) and distributor configuration was maintained, with the number of heater tubes being reduced according to the same 4/10 ratio. The comparison of reactor dimensions between the model and commercial scale is shown in Fig. 6.6.

The conical base section of the TDN reactor was the most complex part to model as it accommodates the central and upper air fluidization rings, as well as the ring to hold the internal heating tubes. A diagram of the lab scale conical section is shown in Fig. 6.7, where the 3 gas injection levels are shown schematically as shaded at levels 1, 3 and 5, whilst the dummy heating tubes (levels 2 and 4) are unshaded in the diagram. Although the geometrical characteristics of the conical base of the real TDN (see Fig. 6.8a) could not be replicated exactly in the scale down model (Fig. 6.8b), the design of the conical base was carefully devised so to obtain the required flow rates. Gas flow rates for the commercial scale and scale down reactors are shown in the Table 6.4.

Having matched the geometric scaling described above, the simplified Glicksman scaling laws allowed determination of the physical characteristics of the bed material and the operating conditions to be adopted so as to achieve fluid dynamically equivalent conditions in the scale down model compared to the full-scale system. Table 6.5 shows a comparison of the operating parameters used in TDN reactor, the values of the calculations for an exact match and the values which were selected to be as close as practicable.

A comparison of the reduced set of reaction parameters matched for hydrodynamic equivalence is shown in Table 6.6. The closest practical solid (to uranium trioxide powder) chosen for the study was (titanium oxide) sand which has a particle density of 4600 kg/m^3 and the desired particle size distribution and gave a

Scale:	10		4	
Dimensions:				
	Real size		Lab scale	
A	3.486	m	1.3944	m
B	1.216	m	0.4864	m
C	0.3	m	0.12	m
D	0.914	m	0.3656	m
E	1.42	m	0.568	m
F	0.76	m	0.304	m
G	4.223	m	1.6892	m

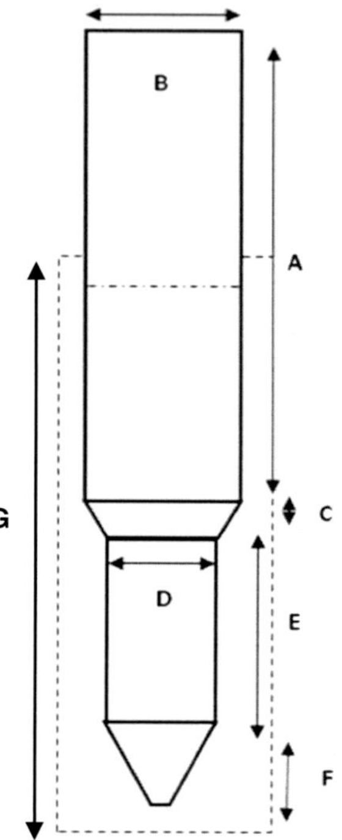

Fig. 6.6 Comparison of dimensions between commercial and lab scale reactor

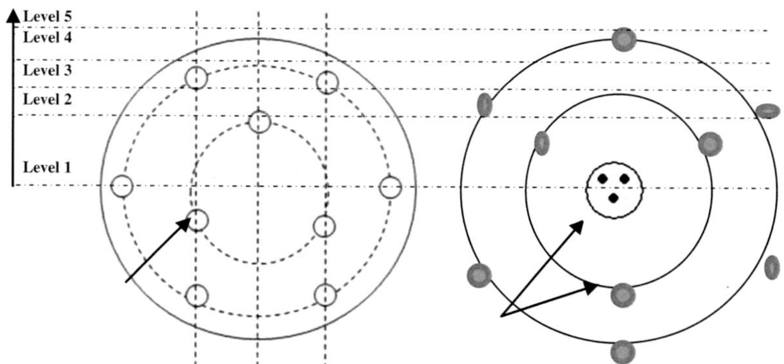

Fig. 6.7 Design of the conical section, comprising the aeration nozzles and the dummy heating pipes

satisfactory match to the 4800 kg/m^3 required by the simulation calculations. Titanium dioxide powder was used for the experiments, which was a close match to the material identified with the scaling rules and also matched the physical characteristics of the uranium trioxide (UO_3) produced in the TDN reactor at Sellafield. Figure 6.8 shows the final CAD design and the 4/10th scale TDN reactor which was designed and built at UCL.

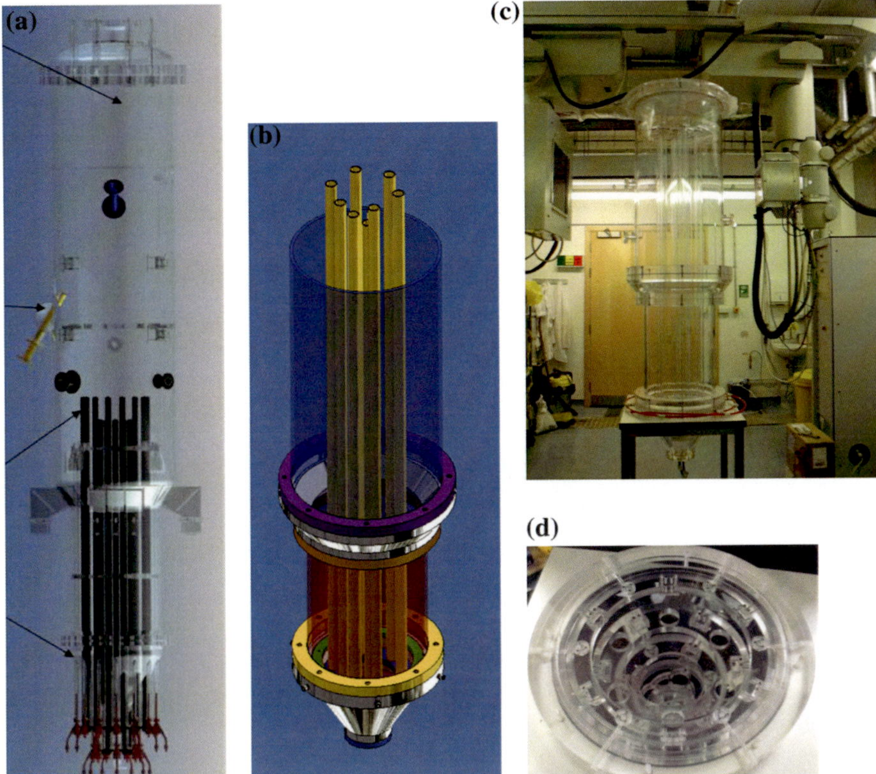

Fig. 6.8 **a** schematic of the TDN reactor at Sellafield; **b** CAD design of Lab Reactor alongside the Perspex Scale down TDN Modeling the X-ray Cell at UCL **c**; and **d** a detailed view of the conical section (refer also to Fig. 6.7)

Table 6.4 Gas flow rates for the commercial scale and scale down reactors

	Original TDN			Cold model		
	Central nozzle	Central ring	Upper ring	Central nozzle	Central ring	Upper ring
Number of air nozzles	7 (fissures)	7	14	3	3	6
Air rate (m^3/h)	90.00	135.00	270.00	10.33	15.5	31.00
Air flow through 1 nozzle (m^3/h)	13.00	19.28	19.28	3.44	5.16	5.16

Although with this particular application, experimental data from the full scale plant could not be provided to validate the application of the scaling rules, the experimental evidence (over 40,000 X-ray images taken) obtained with the 4/10th scaled down TDN model was successfully used to provide information on the jet penetration into the conical section of the TDN, the bubble dynamics evolving in

Table 6.5 Comparison of operating parameters

Parameter	TDN	Exact cold model	Actual cold model
T (C)	300	10	15
P (bar)	3	1	1
μ (kg/ms)	2.993E−05	1.78E−05	1.81E−05
ρ_g (kg/m3)	1.841	1.24	1.33
ρ_s (kg/m3)	7100	4800	4600
Φ	0.77	0.77	0.77
u_{mf} (m/s)	0.00817	0.00513	0.00457
u_o (m/s)	0.26	0.16	0.16
D (m)	0.914	0.3656	0.36
dp (μm)	100	63.1	60

Table 6.6 Simplified scaling parameter values

Scaling parameter	TDN	Cold model
ρ_s/ρ_g	3855.98	3461
u_0^2/gD	0.00756	0.00751
u_0/u_{mf}	31.85	32.01
Φ	0.77	0.77

the upper sections of the reactor, the bubble induced solids mixing and elutriation, and nozzles performance. Figure 6.9 shows some of the X-ray images obtained during this project.

Thanks to the application of the scaling rules for fluidization, extensive and systematic experiments were undertaken in the 4/10 scale TDN and several recommendations were made which led to the improvement of the solids mixing, heat transfer and reactor control of the real TDN. Sellafield Ltd., as a result of this project, has seen a massive improvement in operational reliability and throughput of the Magnox TDN Reactors—these are now no longer perceived as "high risk" to the operation of Magnox Reprocessing. The objectives of the NDA's UK Strategy for hazard reduction have been addressed and the risk of the possible requirement for alternative long term fuel storage for Magnox spent fuel has been mitigated, with potential substantial savings for the UK Taxpayer.

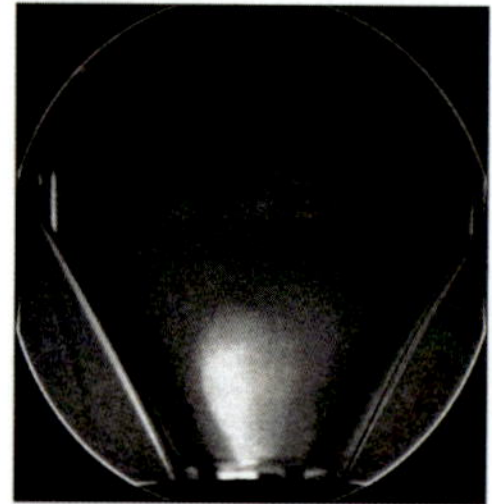

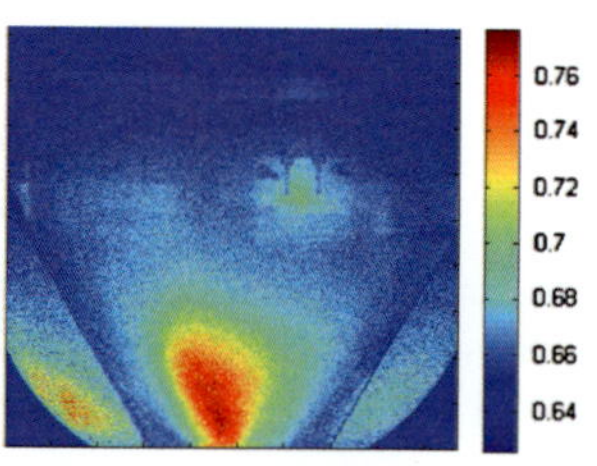

Fig. 6.9 (*left*) central nozzle jet into the conical section of the 4/10 scale TDN; (*centre*) voidage distribution around the central jet; (*right*) particle motion in the freeboard

References

Alliston MG, Wu S (1996) SO_2 distribution in large CFB combustors and its impact on sorbent requirements. CFB Technology. Science Press, Beijing, pp 327–332

Anderson TB, Jackson R (1967) A fluid-mechanical description of a fluidized bed. Ind Eng Chem Fundam 6:527–539

Chang H, Louge M (1992) Fluid dynamic similarity of circulating fluidized beds. Powder Technol 70:259–270

Crowther ME, Whitehead JC (1978) Fluidization of fine powders at elevated pressures. Fluidization. Cambridge University Press, Cambridge

Diks C, van Zwet WR, Takens F, de Goede J (1996) Detecting differences between delay vector distributions. Phys Rev E 53:2169

Di Felice R, Rapagna S, Foscolo PU (1992) Dynamic similarity rules: validity check for bubbling and slugging beds. Powder Tech 71:281

Fitzgerald T, Bushnell D, Crane S, Sheih Y-C (1984) Testing of cold scaled-bed modelling for fluidized-bed combustor. Powder Technol 38:107–120

Foscolo PU, Di Felice R, Gibilaro LG, Pistone L, Piccolo V (1990) Scaling relationships for fluidization: the generalised particle bed model. Chem Eng Sci 45:1647–1651

Gardiner CW (1997) Handbook of stochastic methods, 2nd edn. Springer, Berlin

Gibilaro LG, Hossain I, Foscolo PU (1986) Aggregate behaviour of liquid fluidized beds. Can J Chem Eng 64:931–938

Gibilaro LG (2001) Fluidization-dynamics. Butterworth Heinemann, Oxford

Glicksman LR (1988) Scaling relationships for fluidized beds. Chem Eng Sci 43:1419–1421

Glicksman LR. Hyre M, Woloshun K (1993) Simplified scaling relations for fluidized beds. Powder Technol 77:177–199

Glicksman LR, Hyre MR, Farrell PA (1994) Dynamic similarity in fluidization. Int J Multiph Flow 20:331–386

Glicksman LR (2003) Fluidized-bed scale-up. In: Yang W-C (ed) Handbook of fluidization and fluid-particle systems. Marcel Dekker, New York

Holmes R, Materazzi M, Gallagher B (2015) 3D printing and X-ray imaging applied to thermal denitration at Sellafield. Nucl Future J 12(6):33

Horio M, Nonaka A, Sawa Y (1986) A new similarity rule for fluidized-bed scale-up. AIChE J 32:1466–1482

Horio M, Ishii H, Kobukai Y, Yamanishi N (1989) A scaling law for circulating fluidized beds. J Chem Eng Jpn 22:587–592

Jacob KV, Weimer AW (1987) High temperature particulate expansion and minimum bubbling of fine carbon powders. AIChE J 33:1698–1706

Leckner B, Werther J (2000) Scale-up of circulating fluidized-bed combustion. Energy Fuels 14:1286–1292

Leckner B, Szentannai P, Winter F (2011) Scale-up of fluidized-bed combustion. A review. Fuel 90:2951–2964

Lettieri P (2014) Design, manufacture and X-ray Imaging of a scaled down TDN fluid bed reactor. Keynote, Particulate systems analysis conference, Manchester, 15–17 September

Lettieri P, Yates JG (2013) New Generation X-ray Imaging for multiphase systems. In: Fluidization XIV, Leeuwenhorst, Noordwijkerhout, The Netherlands

Materazzi M, Lettieri P, Holmes R, Gallagher B (2016) Magnox reprocessing TDN reactors: utilising 3D printing and X-ray imaging to re-design and test fluidising air nozzles. In: Proceedings of Waste Management, Phoenix, Arizona

Newton D (2004) Revealing the secrets of fluidized beds, exploiting links between academia and industry. Ingenia, (14):47–52

Nicastro MT, Glicksman LR (1984) Experimental verification of scaling relationships for fluidized beds. Chem Eng Sci 39:1381–1391

Roy R, Davidson JR (1988) Similarity between gas-fluidized beds at elevated temperature and pressure. Fluidization VI. Engineering Foundation, New York, pp 293–300

Rüdisüli M, Schildhauer TJ, Biollaz SMA, van Ommen JR (2012) Scale-up of bubbling fluidized-bed reactors: a review. Powder Technol 217:21–38

Sanderson J, Rhodes M (2005) Bubbling fluidized bed scaling laws: Evaluation at large scales. AIChE J 51:2686–2694

Takens F (1981) Detecting strange attractors in turbulence. In: Rand D, Young L-S (eds) Lecture notes in Mathematics vol 898, Dynamical systems and turbulence. Springer, Berlin

van Ommen JR, Schouten JC, Coppens M-O, van den Bleek (1999) Monitoring fluidization by dynamic pressure analysis. Chem Eng Technol 22:773–779

van Ommen JR, Coppens M-O, van den Bleek CM, Schouten JC (2000) Early warning of agglomeration in fluidized beds by attractor comparison. AIChE J 46:2183–2197

van Ommen JR, Sanderson J, Nijenhuis J, Rhodes MJ, van den Bleek CM (2004) Reliable validation of the simplified scaling rules for fluidized beds. Fluidization XI Engineering Conferences International, New York, pp 451–458

van Ommen JR, Teuling M, Nijenhuis J, van Wachem BGM (2006) Computational validation of the scaling rules for fluidized beds. Powder Technol 163:32–40

Zhang MC, Yang RYK (1987) On the scaling laws for bubbling gas-fluidized-bed dynamics. Powder Technol 51:159–165

Author Index

J.G. Yates and P. Lettieri, *Fluidized-Bed Reactors: Processes and Operating Conditions*, Particle Technology Series 26,
DOI 10.1007/978-3-319-39593-7

Subject Index

J.G. Yates and P. Lettieri, *Fluidized-Bed Reactors: Processes and Operating Conditions*, Particle Technology Series 26,
DOI 10.1007/978-3-319-39593-7

Zeitfracht Medien GmbH
Ferdinand-Jühlke-Straße 7
99095 Erfurt, Deutschland
produktsicherheit@kolibri360.de